ITOGI
Summaries of Scientific Progress

OCEANOLOGY
Volume 1

Scientific Editors

A. P. Kapitsa, P.S. Lineykin, K. S. Losev

GEOPHYSICS SERIES

Editor-in-Chief

A. P. Kapitsa, USSR Academy of Sciences

G.K.HALL & CO.
GKH **Boston, Massachusetts**
1974

Original Russian text published by the All-Union Institute for Scientific and Technical Information (VINITI) in Moscow in 1971 as a volume of Itogi Nauki. Translation published under contract with the USSR Copyright Agency (VAAP).

Printed in the United States of America.

Library of Congress Catalog Card Number 74-8870
ISBN 0-8161-2000-5

ИТОГИ НАУКИ

ОКЕАНОЛОГИЯ
Том 1

Научные редакторы:

А. П. Капица, П. С. Линейкин, К. С. Лосев

СЕРИЯ ГЕОФИЗИКА

Главный редактор:

член-корреспондент АН СССР А. П. Капица

Г. К. Холл и Ко.

GKH Бостон, США

1974

Оригиальный русский текст напечатан Всесоюзным Институтом
Научной и Технической Информации в Москве в 1971 как том серии
«Итоги Науки». Перевод издан по контракту с ВААП-ом.

Число в каталоге Библиотеки Конгресса (США) 74-8870

Число ISBN 0-8161-2000-5

Папечатано в Соединённых Штатах Америки.

CONTENTS

INTRODUCTION

This volume is the first in the Itogi series on Oceanology. It reviews the most significant progress internationally in the two or three years prior to its original publication in Russian in 1971 in such areas as the technology of oceanographic research, air–sea interaction (theoretical models of the boundary layers of the air and sea), tides, ice, and ocean forecasts. It is designed for use by practicing scientists, researchers, and engineers as well as by teachers and students.

Since 1964, the Geophysics Department of the All-Union Institute for Scientific and Technical Information (VINITI) has published reviews on various aspects of geophysics under the general heading, Summaries of Scientific Progress. Now, because of the great variety of topics comprising the overall discipline, this series has been reorganized and will be published as a set of individual, sequential volumes, following the nomenclature of the sections of *Abstracts Journal–Geophysics.* The following volumes of Summaries of Scientific Progress are planned, with each appearing about every two years:

Oceanology, Meteorology and Climatology, Physics of the Earth, Glaciology, and Geomagnetism and the Upper Atmosphere

The reviews of the world literature are prepared by highly qualified specialists in their respective fields: Prof. I. S. Berzon, Prof. S. S. Gaygerov, Prof. A. M. Gusev, Prof. B. L. Dzerdzeyevskiy, Prof. A. D. Dobrovolskiy, Prof. L. I. Dorman, Prof. A. I. Duvanin, Dr. G. S. Ivanov-Kholodniy, Prof. Yu. D. Kalinin, Prof. K. Ya. Kondratyev (Corresponding Member, USSR Academy of Sciences), Ye. A. Koridalin, Prof. A. S. Monin (Corresponding Member, USSR Academy of Sciences), Dr. N. V. Pushkov,

SEA–ATMOSPHERE INTERACTION

(Theoretical Models of the System of Atmospheric and Sea Boundary Layers)

B. A. Kagan

I. INTRODUCTION

In 1963, when the paths that research on sea–air interaction would take were but dimly seen, a group of leading American geophysicists stated [44] : "We are beginning to realize dimly, although our information on this score is far from complete, that the atmosphere and the oceans, which together constitute the fluid portions of the earth, actually function as one huge mechanical and thermodynamical system." And further: "A physical understanding of the processes of air–sea interaction should be one of the major objectives of geophysics during the coming decade."

A decade has not even elapsed since then, but there already is substantial progress in research on sea–air interaction, as evidenced, among others, in a huge outpouring of publications on the various aspects of this problem. By itself this fact does not, of course, mean that the problem is solved, or even close to a solution. However, there is no doubt that there have been advances in understanding of the physical nature of the sea–air interaction.

Special attention was paid in the last few years to the analysis of small-scale atmosphere–sea interaction, including the still largely unknown mechanisms of mass and energy transfer across the air–water interface. Preliminary results of these studies have been reported in hundreds of papers, several reviews [19, 20, 38, 41, 48, 52, 53, 62] and in two monographs [21, 58].

The present review deals with work in one area of the above problem, an area which was not adequately covered by the above-cited reviews and monographs, namely the development of the physical theory of interaction between boundary layers situated on the two sides of the sea–air contact zone.

The basic reason for the existence of boundary layers is the break of continuity of the density at the water–air interface. It is this discontinuity that induces large vertical velocity gradients in the atmosphere near the sea surface and the resultant dynamic instability of the air flow and turbulence. The turbulent shear stress that the moving air flow exerts induces flow of surface water, which in turn entrains water at greater depth. This is how drift currents in the ocean are generated. With depth, the drift current velocity decreases and vanishes at some finite depth called the depth of the friction layer. As a result, large vertical velocity gradients may be induced in the surface layer of the sea, and hence turbulence may be generated.

The intensity of turbulence in both air and water is a function of the vertical velocity gradients and density which, in turn, are themselves turbulence-dependent. Therefore, the velocity and density distributions, the intensity of turbulence and the turbulent momentum, heat, and energy fluxes that are governed by these distributions cannot be assumed to be independent variables, but must be determined together for given sets of background parameters.

For the system of two boundary layers (in the air and the sea), the background parameters are the meteorological elements at the top of the planetary boundary layer of the atmosphere, that is, the geostrophic wind speed, the temperature, the air humidity, and the radiation balance, as well as the hydrological characteristics at the bottom of the sea–surface boundary layer, that is, the geostrophic current velocity, the temperature, and the salinity of the water. It is these factors that are responsible for the boundary layers and determine their structure, conditions at the water–air interface, and the interrelationship between the elements.

The background parameters of the two boundary layers are, in turn, determined by the interactions between the atmospheric boundary layer and free atmosphere and between the sea boundary layer and subsurface water. Since, however, the turbulent fluxes of various substances decrease away from the interface and vanish at the top and the bottom of the respective boundary layers, the boundary layers, free atmosphere, and the subsurface water are not coupled by transport of properties by the turbulent fluxes, but by exchange processes occurring on a completely different scale. Thus, the coupling between the atmospheric boundary layer and free atmosphere, and between the sea boundary layer and the subsurface water involves ordered vertical motion at a much lower rate than that of turbulent fluctuations.

The mutual adaptation time of the velocity fields in the boundary layers of the atmosphere and the sea is much shorter than the adaptation time in the free atmosphere and, of course, very much shorter than deep in the ocean. Therefore, one can, under some circumstances, consider the dynamics of the boundary layers separately from those of the air system as a whole, without worrying about feedback mechanisms. In this sense, one can assume that the system consisting of the air and sea boundary layers is to some extent independent.

II. THEORETICAL MODELS OF THE SYSTEM OF AIR-SEA BOUNDARY LAYERS, BASED ON *A PRIORI* ASSUMPTION ABOUT THE MAGNITUDE AND PROFILE OF THE TURBULENCE COEFFICIENT

In analyzing the wind pattern and ice motion during the drift of the *Fram*, Nansen found that the ice drift in the Arctic Ocean deviated 30°-40° to the right of the wind direction. He explained this by the effect of rotation of the Earth, and speculated that since the surface water is ice-driven, the flow thereby generated in the ocean should deviate even more from the wind direction.

The theory of this problem was then analyzed by Ekman [45] who in 1902 obtained results that agreed with Nansen's qualitative conclusions. Somewhat later, Ekman confirmed by theoretical analysis the clockwise rotation (in the (northern hemisphere) of the drift current velocity vector with depth [46]. In this same paper Ekman also speculated on the possibility that his results may be applicable to the description of the change in wind speed with height over land. This idea was realized later by Åkerblom [43] and, independently, by Exner [47].

We shall consider at some length the Exner paper, which was the first to treat the air and sea boundary layers as a unified dynamic system. In the steady-state case of rectilinear and uniformly distributed isobars, the flow in the air and sea boundary layers can be described by

$$\frac{d}{dz_i} k_i \frac{du_i}{dz_i} + f(v_i - V_{gi}) = 0; \quad (1)$$

$$\frac{d}{dz_i} k_i \frac{dv_i}{dz_i} - f(u_i - U_{gi}) = 0; \quad (2)$$

where u_i, v_i are the velocity components of the wind or drift current along the x and y axes; $f=2\omega\sin\varphi$ is the Coriolis parameter; z_i is the vertical axis, pointing up in the atmosphere and down in the sea (the origin being at the free surface); the subscripts i take the values 1 and 2, i = 1 and i = 2 relating to the air and sea boundary layers, respectively; U_{gi} and V_{gi} are the geostrophic wind or geostrophic current components:

$$U_{gi}=-\frac{1}{f\rho_i}\frac{\partial p_i}{\partial y};\quad V_{gi}=\frac{1}{f\rho_i}\frac{\partial p_i}{\partial x}$$

(p_i and ρ_i are the pressure and mean density of air or water).

Exner solved Eqs. (1) and (2) under the assumption that U_{g2}, $V_{g2}=0$, the eddy viscosity coefficient k is constant with height and is given, and that the following conditions are satisfied at the water-air interface:

$$u_1=u_2;\ v_1=v_2;\ k_1\rho_1\frac{du_1}{dz_1}=-k_2\rho_2\frac{du_2}{dz_2};\ k_1\rho_1\frac{dv_1}{dz_1}=-k_2\rho_2\frac{dv_2}{dz_2}$$

$$\text{at}\ \ z_i=0; \tag{3}$$

$$u_i\to U_{gi};\ v_i\to V_{gi}\ \ \text{at}\ \ z_i\to\infty. \tag{4}$$

Conditions (3) mean continuity of the velocity and the momentum fluxes on the free surface of the sea, while condition (4) denotes the limitation on velocity at great heights within the corresponding boundary layers.

The final expressions for the distribution of wind speed or current are

$$\begin{aligned}u_i&=U_{gi}+e^{-a_i z_i}\left[(u_0-U_{gi})\cos a_i z_i+(v_0-V_{gi})\sin a_i z_i\right];\\ v_i&=V_{gi}+e^{-a_i z_i}\left[(v_0-V_{gi})\cos a_i z_i-(u_0-U_{gi})\sin a_i z_i\right],\end{aligned} \tag{5}$$

the current velocity components on the ocean surface being given by

$$u_0=\frac{U_{g1}}{1+\frac{\rho_2}{\rho_1}\sqrt{\frac{a_1}{a_2}}};\quad v_0=\frac{V_{g1}}{1+\frac{\rho_2}{\rho_1}\sqrt{\frac{a_1}{a_2}}}; \tag{6}$$

where $a_i=\sqrt{\frac{\omega\sin\varphi}{k_i}}$.

Using the turbulence coefficient estimates from [59], Koshmider has calculated the vertical variations of the wind and the drift current, and found that the velocity distribution in the air and sea boundary layers is described by Ekman's double spiral (Fig. 1).

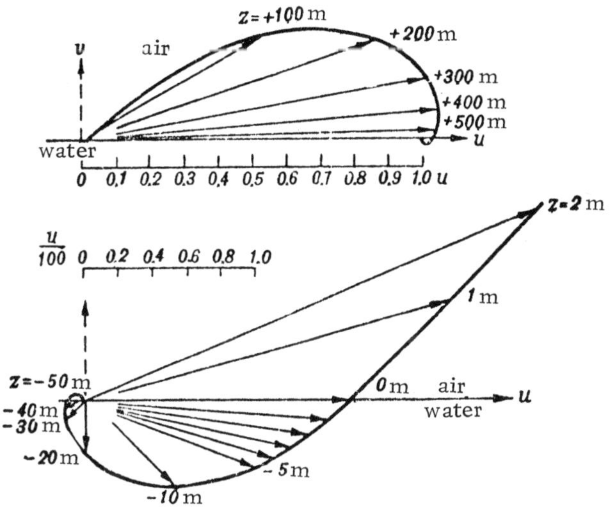

FIGURE 1. EKMAN'S DOUBLE SPIRAL ACCORDING TO KOSHMIDER [26]. THE LOWER SPIRAL IS MAGNIFIED BY A FACTOR OF 50; $z_1 = z$; $z_2 = -z$.

A nonzero but very small wind velocity was found at the water surface. Its direction coincided with that of the geostrophic wind, while its magnitude was the same as that of the surface current. The conclusions as to the counterclockwise rotation of the wind with altitude in the surface atmospheric layer was derived from this fact.

Exner's theory was further advanced by Shvets [42] and Hesselberg [50]. To obtain a better fit between the calculated and measured wind speeds, these authors allowed for vertical variations in the atmospheric pressure gradients. However, unlike Exner and Shvets, Hesselberg assumed slip as the boundary condition at the sea surface. Both Shvets and Hesselberg still assumed the eddy viscosity coefficient to be height-independent and given.

With these conditions, the computed vertical variations in the absolute wind speed qualitatively approached the real fluctuations. Thus, in the first few hundred meters above the surface, the wind speed increases very rapidly. The increase with height is slower between 500 and 1000 m or ceases completely, and the wind picks up again at greater heights.

Laykhtman [28] has generalized Exner's theory to the case in which the sea is covered with free ice floes. In this case, the condition of continuity of momentum flux was replaced by the condition of the balance of forces acting on an ice volume of unit cross section and mass M. In projection on the horizontal coordinate axes, this condition can be written as

$$
\begin{aligned}
&fMv_0+\rho_1 k_1 \frac{du_1}{dz_1}\bigg|_{z_1=z_0} + \rho_2 k_2 \frac{du_2}{dz_2}\bigg|_{z_2=z_{02}} = 0;\\
&-fMv_0+\rho_1 k_1 \frac{dv_1}{dz_1}\bigg|_{z_1=z_{01}} + \rho_2 k_2 \frac{dv_2}{dz_2}\bigg|_{z_2=z_{02}} = 0;
\end{aligned}
\tag{7}
$$

where z_{01} and z_{02} are the roughnesses of the upper and lower ice edges, respectively.

If k_i is assumed to be independent of height, then z_{0i} may be considered to equal zero and then the expressions for the velocity components of the ice drift are

$$
\begin{aligned}
u_0 &= 2U_{g1}\frac{1+Ma_2/\rho_2}{\frac{\rho_2}{\rho_1}\sqrt{\frac{a_1}{a_2}}\,[1+(1+2Ma_2/\rho_2)^2]};\\
v_0 &= -2U_{g1}\frac{Ma_2/\rho_2}{\frac{\rho_2}{\rho_1}\sqrt{\frac{a_1}{a_2}}\,[1+(1+2Ma_2/\rho_2)^2]}.
\end{aligned}
\tag{8}
$$

In deriving (8), the x axis was assumed to point in the direction of the geostrophic wind; it was also assummed that $|u_0|$, $|v_0| \ll U_{g1}$.

According to (8), $u_0>0$, $v_0<0$. This means that over deep sea the ice always drifts toward high pressures, and in the presence of floes the counterclockwise wind shear in the lower atmosphere will exceed the value predicted by Exner.

III. SIMPLEST CLOSED MODELS OF THE AIR-SEA BOUNDARY-LAYER SYSTEM

1. Horizontally homogeneous air and sea boundary layers

The systems of equations employed in the papers cited above are not closed. Apart from the wind-speed and drift-velocity components, the system contains two more unknown functions, namely the turbulence coefficients for both boundary layers. For closure, the equations of motion for the two boundary layers must be complemented with two expressions relating the wind-speed distribution to the turbulence coefficients. This was done by Laykhtman [27, 29] by means of the integral form of the balance equation for turbulence energy:

$$
\int_0^{H_i} k_i\left[\left(\frac{du_i}{dz_i}\right)^2+\left(\frac{dv_i}{dz_i}\right)^2 - g\frac{d\ln S_i}{dz_i}\right]dz_i - \langle \varepsilon_i \rangle H_i = 0, \tag{9}
$$

where g is the acceleration of gravity, $S_i = \Theta$ is the potential temperature in the air boundary layer, $S_2 - \rho_2$ is the same in the sea boundary layer.

In writing the balance equation in form (9), we assume that diffusion fluxes vanish on the outer limits of the boundary layers and at the air–water interface. While this equation does have obvious advantages in computing the turbulence coefficients, it also has drawbacks, namely the need to find two new unknown parameters, that is, the mean rate of turbulent-energy dissipation $\langle \varepsilon_i \rangle$ and the height of boundary layer H_i. In addition, one must approximate the vertical profile of the turbulence coefficient by means of a known function of the height.

Two methods for determining $\langle \varepsilon_i \rangle$ are now most widely used [33]. The first of these assumes that the dissipation is proportional to the turbulent-energy input due to the kinetic energy of the mean flow. The proportionality factor

$$\delta_i = \frac{\langle \varepsilon_i \rangle H_i}{\int\limits_0^{H_i} k_i \left[\left(\frac{du_i}{dz_i}\right)^2 + \left(\frac{dv_i}{dz_i}\right)^2 - g\,\frac{d \ln S_i}{dz_i} \right] dz_j} \tag{10'}$$

is [12, 30] about 0.3.

The second method for determining $\langle \varepsilon_i \rangle$ assumes that $\langle \varepsilon_i \rangle$ is a function of the boundary layer height and the turbulence coefficient, averaged over the boundary layer. The form of this function can be found by dimensional analysis:

$$\langle \varepsilon_i \rangle = \bar{c}\,\frac{\bar{k}_i^3}{H_i^4}, \tag{10''}$$

where $\bar{c}$ is a numerical factor derived from empirical data.

The boundary-layer height H_i can be found either by assuming that it is equal to the height at which the derivative of the velocity modulus vanishes for the first time

$$\frac{d}{dz_i}\,(u_i^2 + v_i^2)^{1/2}\Big|_{z_i = H_i} = 0, \tag{11'}$$

or that it is equal to the height at which the Ekman condition

$$[(u_i - U_{gi})^2 + (v_i - V_{gi})^2]^{1/2}_{z_i = H_i} = e^{-\pi}(U_{gi}^2 + V_{gi}^2)^{1/2}. \tag{11''}$$

is satisfied.

If there are no geostrophic currents in the water boundary layer, (11'') can be rewritten as

$$(u_2^2 + v_2^2)^{1/2}\Big|_{z_2=H_2} = e^{-\pi}(u_0^2 + v_0^2)^{1/2}. \qquad (11''')$$

The system of Eqs. (1), (2), (9)–(11) is closed.

This system of equations, with k_i = const, is solved in [7]. The formulas derived therein for the turbulence properties of the air and sea boundary layers are

$$H_1 = -\frac{T_H - T_0}{2\gamma_a} + \left[\left(\frac{T_H - T_0}{2\gamma_a}\right)^2 + (1-\delta_1)\frac{2.28G_1^2}{\frac{g}{\theta}\gamma_a}\right]^{1/2};$$

$$k_1 = \frac{fH_1^2}{10.4};$$

$$k_2 = \frac{\rho_1}{\rho_2} G_1 \left[\frac{(1-\delta_2)}{2\pi}\,\frac{fk_1}{\frac{g}{\rho_2}\left\langle\frac{d\rho_2}{dz_2}\right\rangle}\right]^{1/2}; \qquad (12)$$

$$H_2 = \pi\sqrt{\frac{2k_2}{f}},$$

where $G_1 = \sqrt{U_{g1}^2 + V_{g1}^2}$ is the geostrophic wind-speed modulus, T_H and T_0 are the temperatures at the top of the air boundary layer and at the sea surface, respectively, γ_a is the adiabatic lapse rate, $\left\langle\frac{d\rho_2}{dz_2}\right\rangle$ is the mean density gradient in the water boundary layer.

Equations (12) are very simple and may be used to explain the dependence of the desired functions on background variables that control them. In particular, analysis of (12) shows that the sea turbulence coefficient increases with the wind speed, but that this dependence is not quadratic (as was assumed earlier) but is only $k_2 \sim G_1^{3/2}$. Furthermore, the wind coefficient c_0/G_1 (here c_0 is the surface current velocity modulus) is a function of not only wind speed ($c_0/G_1 \sim G_1^{1/4}$), but also of the stability of the air and the sea.

Of equal interest is the conclusion about the latitudinal dependence of the depth of the friction layer in the sea [9].

From (12), it would appear that $H_2 \sim f^{1/4}G_1^{3/2}$. But analysis of actual data has shown that the geostrophic wind speed increases with latitude. From this follows a conclusion that contradicts the commonly accepted view that the depth of the friction layer decreases from equator to pole.

The above model gave a qualitatively correct description of the real qualitative relationships that exist between the background parameters and the various characteristics of the system of boundary layers. But, because the model uses turbulence coefficients that are height-invariant, it failed to produce a quantitative agreement between the calculated results and observations. Therefore, further work on the improvements of the model tended toward replacement of the height-invariant turbulence coefficients with functions of vertical coordinates. For instance, Eqs. (1), (2), (9)–(11) were solved in [8] on the assumption that the atmospheric turbulence coefficient can be approximated by

$$k_1 = \begin{cases} k_{11} z, & \text{at} \quad z \leqslant h \\ k_{11} h, & \text{at} \quad z > h, \end{cases} \tag{13}$$

while also assuming that the turbulence coefficient in the sea is constant with depth. Then, Eq. (9), specifying the turbulent energy balance, was used to determine the height of the inflection h, while the limiting Prandtl relation

$$k_{11} = \frac{\varkappa^2 c_1}{\ln z_{11}/z_{01}} \tag{14}$$

(where c_1 is the absolute wind speed at $z_1 = z_{11}$, z_{01} is the roughness of the water surface, and $\varkappa$ is the von Karman constant) was used to define the turbulence coefficient at $z_1 = z_{11}$.

Other possible models of variations of air and sea turbulence coefficients were examined in [11], namely the case in which the distribution of the air turbulence coefficient is given by model (13) and where the sea turbulence coefficient is a linear function of the depth. An analysis of the effect of the nature of variation of the air and sea turbulence coefficients on the vertical structure of the boundary layers has shown that the replacement of a vertically varying turbulence coefficient with some mean value has no effect on the calculated wind-speed profile in the air boundary layer and changes but slightly the calculated drift current distribution in the sea. This change shows up as a small current velocity increase in the upper sea layer and as a velocity decrease in the lower part of the friction layer.

If the turbulence coefficient in the atmospheric boundary layer is assumed constant, then the computed drift current velocity increases, but not so abruptly as might have been expected (by not more than 60%).

It appears that the angle between the surface current vector and ground isobar is fairly sensitive to the mode variation of the turbulence coefficient. Whereas with a height-invariant turbulence coefficient the surface current and geostrophic wind vector practically coincide, with a variable coefficient the surface current flows toward higher pressures and makes an angle of about 30° with the ground isobar.

To find the optimum approximation of the vertical profile of the turbulence coefficient in the air and sea boundary layers, the results of the wind-speed and current calculations were compared with observations. However, since observational data, obtained under controlled conditions, were extremely scant, the comparison was made with the empirical estimates of wind coefficient listed in Table 1.

Table 1

Values of wind coefficient as reported in the literature

Author	Wind coefficient	Author	Wind coefficient
Hela	0.0092	Rossby and Montgomery	0.0169
Pulman	0.0092	Dieser	0.0181
Torade	0.0092	Stommel	0.0193
Wiedeman	0.0107	Chatley	0.0200
Dinklage	0.0107	Tikner	0.0200
Witting	0.0107	Yeo	0.0200
Struisky	0.0140	Callaghan	0.0221
Ekman	0.0147	Francis	0.0221
Scraton	0.0200	Van Dorn	0.0241
Stevenson	0.0134–0.0200	Tomczak	0.0294
Munk	0.0167–0.0200	Galles	0.0295
		Moon	0.328

Note: Table taken from [35].

It can be seen from the Table that there is no agreement on the value of the wind coefficient. The scatter of individual values is usually explained by measurement errors. However, while there may be some error in the observations, the data of Table 1 differ because they were obtained from observations of drift of different bodies and, therefore, do not describe the motion of the surface layer itself, but of a layer of a finite depth, which varied from case to case. Second, the wind speed used in calculating the wind coefficient was measured at different heights (from 5 to 20 m) over the sea. Furthermore, measurements of wind speed from a ship at sea are not reliable because the wind pattern is distorted by the ship's hull. The wind coefficients reported by Hughes [51] and Tomczak [61] are probably the most representative. They determined the wind coefficient as the ratio of the current velocity, measured directly on the surface (using post-cards drifting in a layer several millimeters thick), to the geostrophic wind speed, thereby eliminating at least three of the errors listed above.

Allowing for this, let us now proceed to comparison of the wind coefficients, calculated for the various modes of variation of the turbulence coefficients, with the observations. It was shown in [11] that the best fit of the calculated results ($c_0/G_1 = 0.022$) to observations ($c_0/G_1 = 0.0192–0.0226$) is obtained if the atmospheric-turbulence coefficient profile is approximated by an inflected curve and the sea turbulence coefficient is specified by a linear function of depth. All other conditions being equal, replacement of the vertically varying turbulence coefficient by its mean value results in only a small increase in the wind coefficient. In this case $c_0/G_1 = 0.025$.

Thus, in terms of the simple model of the system of air and sea boundary layers, the optimal mode of variations in the turbulence coefficient is one in which k_1 is approximated by Eq. (13) and k_2 is assumed to be constant. The value of k_2, as well as the values of the parameters k_{11} and h in (13), must be determined in each individual case as a function of the background parameters of the system.

The above mode of variations of turbulence coefficients was used to construct a quasistationary model of thermal and dynamic interaction between air and sea boundary layers [14, 15]. In this model, the set of equations of dynamics was completed with heat-input equations for the boundary layers and with the corresponding boundary conditions. The sea-surface temperature was assumed unknown and was computed in the problem from given data on the radiation balance and the temperatures of water and air at the bottom of the sea boundary layer and the top of the planetary atmospheric boundary layer, respectively.

The resultant [35] map of horizontal distribution of water surface temperature in the North Atlantic (see Fig. 2) agrees well with the known current pattern in this area of the ocean. The systematically lower-than-observed calculated temperatures are most probably due to the assumption of a constant turbulent coefficient in the sea boundary layer. Indeed, this would result in higher heat exchange between the sea surface and deeper water, which should

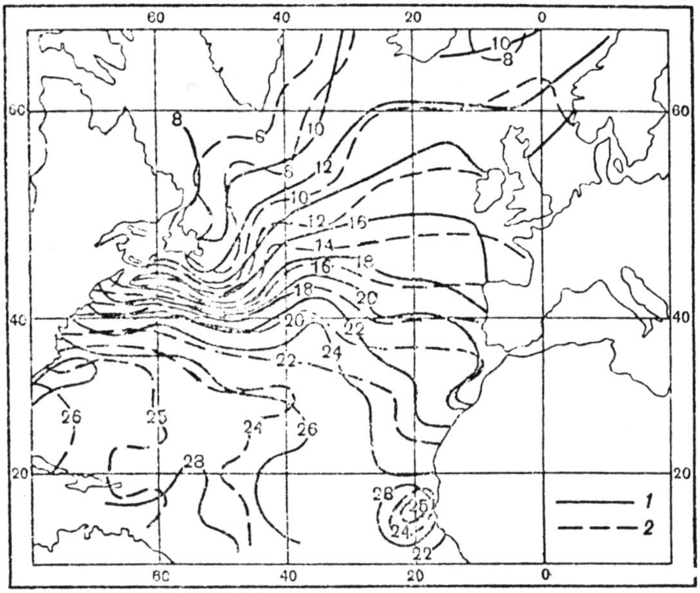

FIGURE 2. DISTRIBUTION OF SURFACE WATER TEMPERATURE IN THE NORTH ATLANTIC, ACCORDING TO LAYKHTMAN ET AL. [35]: 1—CALCULATED; 2—OBSERVED.

result in a lower computed surface temperature for the summer period.

The same model was used [5] to calculate the vertical flow in the friction layer in the North Atlantic. The results obtained add to our knowledge and, besides, are interesting in that they are indicative of a great importance of the horizontal inhomogeneity of the eddy field in generating vertical flow in the upper layer of the ocean.

A very similar and basically simpler model of the sea–air interaction was suggested by Yoshihara [63], who approximated the vertical dependence of the turbulence coefficient for the air by the Monin–Obukhov formula and by the Munk–Anderson formula for the sea. To avoid having to solve the equations of dynamics of the upper ocean layer, Yoshihara assumes that the thermocline has a level in which the vertical velocity gradient exhibits a minimum. At his level, the Richardson number is assumed to equal its maximum value ($R_1 = 3/5$). The turbulence coefficient on the sea surface is assumed to be given.

2. Horizontally inhomogeneous air and sea boundary layers

In the models discussed above, the boundary layers of the atmosphere and the sea were assumed to be horizontally homogeneous.

Much available evidence indicates, however, that in areas of atmospheric and hydrologic fronts the horizontal gradients of all properties can be very much different from zero. For example, in the area of hydrological front in the Sargasso Sea, the temperature gradient is at least one degree/10 km, and there are areas where this gradient is as high as 3° C/100 m.

The first model of interaction between the atmospheric boundary layer and the active ocean layer, allowing for horizontal inhomogeneity of fields of meteorologic and hydrologic variables, was proposed in [17]. This paper gave a solution of a system of steady-state equations, consisting of the equations of flow, state, heat input, and eddy energy balance in both air and sea, together with the equations for diffusion of moisture in air and of salt in the sea. The terms describing advective transport were retained in the equations for heat input and salt diffusion, the horizontal gradients of temperature and salinity being assumed given and constant along the vertical. It was specified that velocities, the temperatures, and the momentum fluxes match at the sea surface. The surface temperature was determined from the heat-balance equation. It was also specified that at the top of the atmospheric boundary layer and at the bottom of the active ocean layer, respectively, the wind and current should become geostrophic, while the temperature and humidity of air as well as the temperature and salinity of water should assume the given values. The problem was solved under the assumption of constant coefficients of turbulence in both air and water.

The expressions obtained for the profiles of wind speed, temperature, and humidity in the atmospheric boundary layer, for temperature and salinity in the active ocean layer, and for the turbulence coefficients and momentum, heat, and moisture fluxes have made it possible to estimate quantitatively the effect of variations in the individual background parameters on the structure of the air–sea system.

The model proposed in [17] was later generalized in [10] for the case of water surface covered with floating ice. The solution decribed above was successfully used to explain some features of the thermal and dynamic interaction between the atmosphere and the ocean in the presence of ice cover. It was found, in particular, that it is the ice cover that causes a decrease in the upward heat flow to the active surface during the cold season. This results in a decrease in the surface water temperature and intensification of atmospheric inversion. However, because of idealization of real processes and the assumption of height-invariant turbulence coefficients, this explanation proved merely qualitative. It was shown in [13] that at k_1 = const the calculated temperature of the upper edge of ice cover is too high in winter, while in summer, when this temperature is equal to that of the melting ice, the calculated heat flow for melting ice* is much too low (by as much as 700%). Therefore, in the improved model of thermal and dynamic interaction between the sea, ice, and atmosphere [13], the vertical profile of the atmospheric turbulent exchange coefficient was approximated by (13). Comparison of the computed results with measurements showed that this model yields fairly accurate data on the temperature of the active surface, the drift velocity, and the rate of ice melting (or accretion). This can be seen from the analysis of the data in Table 2, and from the comparison between the calculated and observed values of melting and drift velocity of ice. The observed mean rate of melting and the mean drift velocity of ice in the Central Arctic are 30–50 and 8 cm per second, respectively, whereas the corresponding calculated values are 57 and 10 cm/sec.

Table 2

Comparison between the calculated and observed temperatures of active ice surface in the Central Arctic [13]

Month	Temperature of active surface	
	Calculated	Observed
January	−31.6	−38.0
February	−32.6	−35.0
March	−32.0	−30.0
April	−19.9	−22.0
May	− 7.5	−10.0
June	0	0
July	0	0
August	0	0
September	− 9.7	− 5.0
October	−18.1	−15.0
November	−28.0	−28.0
December	−30.0	−34.0

*The rate of melting (or ice accretion) was determined from the equation of the heat balance of the lower edge of the ice cover.

Some idea about the potential of the model is given by Figs. 3 and 4, showing the annual pattern of behavior of the main characteristics of the atmospheric boundary layer—ice cover—active sea layer system, as well as the heat balance at the active surface and the rate of snow and ice cover accretion (or melting).

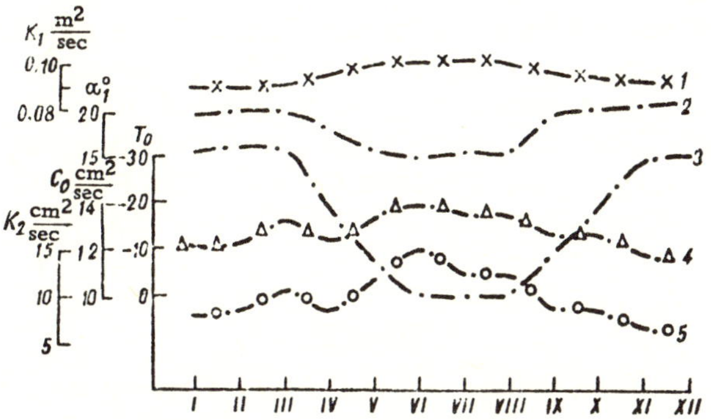

FIGURE 3. THE ANNUAL PATTERNS OF SOME CHARACTERISTICS OF THE SEA-ICE-ATMOSPHERE SYSTEM IN THE CENTRAL ARCTIC, ACCORDING TO KAGAN [13]:

1—TURBULENT EXCHANGE COEFFICIENT k_{11} IN AIR AT THE LEVEL $z_1 = 1$ m; 2—ANGLE α_1 BETWEEN THE ICE DRIFT DIRECTION AND NEAR-GROUND ISOBAR; 3—TEMPERATURE OF ACTIVE SURFACE T_0; 4—ABSOLUTE VALUE OF ICE DRIFT VELOCITY c_0; 5—TURBULENT EXCHANGE COEFFICIENT k_2 IN THE SUBICE-SEA BOUNDARY LAYER.

A different interaction model was examined by Pandolfo [55]. In contrast to [17], the Pandolfo model was intended to study short-period changes in the characteristics of the boundary-layer system. Therefore Pandolfo retains terms describing the time dependence of the characteristics. The horizontal gradients of atmospheric temperature and humidity and of salinity and temperature. in the sea are assumed given (just as in [17]), but in terms of arbitrary functions of vertical coordinates. The turbulence conditions in the air and sea are described in different terms than in [17]. Thus the eddy exchange coefficient for both boundary layers was determined from the modified Prandtl formula rather than from the equations of the turbulent energy balance. Given a stable stratification in both air and sea and the presence of wind

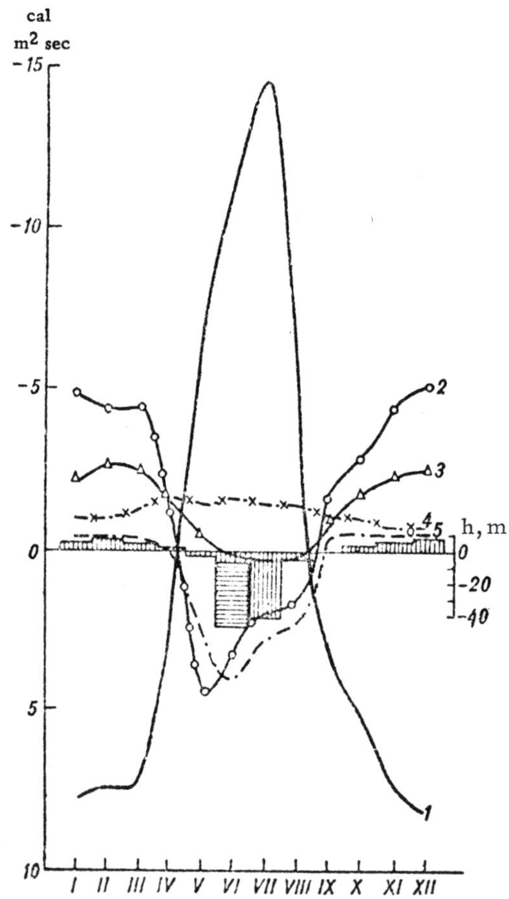

FIGURE 4. THE ANNUAL PATTERNS OF THE COMPONENTS OF HEAT BALANCE OF THE ACTIVE SURFACE IN THE CENTRAL ARCTIC, ACCORDING TO KAGAN [13]:

1—RADIATION BALANCE; 2—TURBULENT HEAT EXCHANGE OF ICE SURFACE WITH ATMOSPHERE; 3—HEAT EXCHANGE BETWEEN THE TOP AND BOTTOM ICE EDGES; 4—TURBULENT HEAT EXCHANGE BETWEEN THE BOTTOM EDGE OF THE ICE AND DEEP-WATER LAYERS; 5—HEAT LOSS BY EVAPORATION; THE VERTICAL (OR HORIZONTAL) HACHURES SHOW THE ACCRETION (OR MELTING) OF ICE AND SNOW.

waves on the free surface, the modified Prandtl formula has the form

$$k_1 = \varkappa^2 \left(z_1 + \frac{\delta_w \lambda}{2} \right) \left(\left| \frac{\partial V_1}{\partial z_1} \right| + S_w \right) (1 + \alpha_0 \mathrm{Ri})^2;$$
$$k_2 = \varkappa_1^2 \left(\left| \frac{\partial V_2}{\partial z_2} \right| + S_w \right) (1 + 10\,\mathrm{Ri})^{-1/2}, \tag{15}$$

where **V** is the velocity vector of wind (V_1) or current (V_2); S_w is the orbital velocity shear of of the wind waves; δ_w and λ are the wave steepness and length, while α_0 and $\varkappa_1^2$ are dimensionless constants equal to 3.0 and 0.02, respectively.

The Pandolfo model equations can be integrated numerically by an implicit scheme which has second-order accuracy in time. The air and sea boundary layers (assumed to be 1500 and 400 m thick, respectively) are subdivided into 30 levels, the vertical interval in the computational grid being variable and decreasing near the air–water interface. The computational time interval was 3 min.

This model was used to calculate the temporal variations of a set of meteorological variables at the standard ship measurement level, of temperature and salinity at the surface, and of current velocity at two points in the North Atlantic [55, 56], namely the point at 13°N 56°W (the BOMEX exploration area) and the site of the weather ship *E*.

In calculating the air and sea turbulent exchange coefficients by relation (15) one assumes that the turbulence in both air and water is sustained by both the mean wind energy and the energy of wind waves. However, we should note here that the "wave" turbulence mechanism becomes effective only in case of very high-wind waves, while at normal conditions, as was shown by Kitaygorodskiy [21], the turbulent energy input induced by orbital motion is negligibly small. In this case $S_w \ll \left| \frac{\partial V_1}{\partial z_1} \right|$.

In addition, the use of (15) in the unsteady-state model of air–sea interaction implicitly assumes instantaneous adaptation of turbulence to the variations in the mean flow. Now, it was shown in [16] that the real adaptation time of the drift-current turbulence is indeed much shorter than the period of variation of the mean velocity. Therefore, in the case of comparatively slow variations in the background parameters one can assume a quasistationary turbulence in the active layer of the sea. However, in the analysis of the short-period fluctuations of meteorological and hydrological variables such assumptions will result in a distortion of the amplitudes and, especially, phase angles of such fluctuations.

Lifshits [36] proposed a simple model of dynamical interaction between the air and sea boundary layers, allowing for unsteady turbulence in both media.

IV. SEMIEMPIRICAL MODELS OF THE SYSTEM OF AIR AND SEA BOUNDARY LAYERS

One feature common to all the models discussed above is the fact that they were derived under the assumption of some approximation of the vertical profiles of the turbulence coefficients in the air and sea. But it was shown in a review by Zilitinkevich, Laykhtman, and Monin [6] that, given the very strong variability of the turbulence coefficient in the atmospheric boundary layer, one cannot reliably describe that layer under all the possible conditions even if one uses adequate (in the sense of yielding realistic data) models of variation of the turbulence coefficient with height.

A similar situation holds for the sea boundary layer, the only difference being that data on the turbulence coefficient are even scarcer. The literature reports very few attempts to determine the turbulent exchange coefficient from direct measurements of the turbulent friction stress and heat flux [24, 25, 39, 49]. We know, however [39], that the turbulent exchange coefficients for momentum and heat peak at depths of the order of 10 to 15 m. Below this level, as can be inferred from the data of Gongwer and Finkle [49], the turbulent exchange intensity decreases sharply with depth and becomes virtually constant from about 100 to 150 m down to 400 m, its magnitude at the latter depths being an order smaller than in the surface layer (depth of about 15 m).

Some additional information on the vertical profile of the turbulence coefficient in the sea might be obtained from the indirect calculation methods proposed by Ekman, Fieldstad, Hidaka and Momoi, and others. However, all these methods, which are based on various theories about the transport mechanism in the sea, require special, very much detailed measurements of the current velocity in the surface layer of the sea. Such data are virtually nonexistent. The more available, but still sporadic, measurements of velocity at standard depth can only be used to estimate the mean turbulence coefficient within the sea boundary layer. The published values of this coefficient are summarized in [54] and show that the eddy viscosity coefficient may range over two orders of magnitude (from 10 to 10^3 cm^2/sec), depending on conditions.

As far as the available techniques for indirect calculation of the turbulent-heat-exchange coefficient are concerned, they do not apply to the top layer of the sea whose vertical structure is quasi-homogeneous.

Thus, the data now available permit only an estimation of the order of magnitude of the turbulence coefficient in the sea boundary layer and the derivation of its approximate profile. For this reason, even elegant and simple models in which the turbulence coefficient is assumed to be a given function of height leave one with a certain feeling of dissatisfaction.

Laykhtman [31] derived a theoretical model of the atmospheric and sea boundary layers without any *a priori* assumptions about the magnitude and profile of the turbulence coefficient. We shall dwell on this model at somewhat greater length not only because it is the best of the existing models, but also to explain the very elegant and ingenious way in which Laykhtman solved the two-layer problem.

If the set of flow equations is closed by means of the turbulent-energy balance and the similarity hypotheses commonly used in the semiempirical theory of turbulence [6], then we shall have a set of equations describing the air and sea boundary layer:

$$\begin{gathered}\frac{d}{dz_i}k_i\frac{du_i}{dz_i}+f(v_i-V_{gi})=0;\\ \frac{d}{dz_i}k_i\frac{dv_i}{dz_i}-f(u_i-U_{gi})=0;\\ k_i\left[\left(\frac{du_i}{dz_i}\right)^2+\left(\frac{dv_i}{dz_i}\right)^2-g\frac{d\ln S_i}{dz_i}\right]-c\frac{b_i^2}{k_i}+\alpha_b\frac{d}{dz_i}k_i\frac{db_i}{dz_i}=0;\\ k_i=b_i\left(\frac{\varkappa\sqrt{c}}{v_{*i}}z_{0i}+\varkappa c^{1/4}\int\limits_{z_{0i}}^{z_i}\frac{dz_i}{\sqrt{b_i}}\right).\end{gathered}\qquad(16)$$

The last expression in (16) was obtained by transforming the Kolmogorov relation for the turbulence coefficient and the generalized Karman formula that was proposed in [6]. Apart from already known nomenclature, (16) contains the following symbols: b_i is the kinetic energy of turbulent fluctuations, v_{*i} is the dynamic velocity, and α_b and c are universal constants equal to 0.73 and 0.046, respectively.

The turbulent-energy balance [the third equation in (16)] contains the function S_i, which describes, as mentioned above, the potential air temperature Θ (at $i=1$) or the sea-water density ρ_2 (at $i=2$). To avoid unnecessary complications associated with the need of finding a new unknown function, one can use the assumption [33] that $k_i\frac{d\ln S_i}{dz_i}=\text{const}$ or a simple interpolation formula for dS_i/dz_i. In both cases the condition $dS_i/dz_i\neq0$ means that the vertical structures of the boundary layers described by (16) are dependent on the nature of stratification in the two adjoining media.

System (16) can be solved with the following boundary conditions:

$$u_i \to U_{gi};\ v_i \to V_{gi};\ b_i \to 0 \ \text{ at } \ z_i \to \infty; \tag{17}$$

$$k_1\rho_1 \frac{du_1}{dz_1} = -k_2\rho_2 \frac{du_2}{dz_2};\ k_1\rho_1 \frac{dv_1}{dz_1} = -k_2\rho_2 \frac{dv_2}{dz_2};$$
$$u_1 = u_2;\ v_1 = v_2;\ b_i = c^{-1/2} v_{*i}^2\ (i = 1, 2). \tag{18}$$

Note that the last condition in (18) follows from the turbulent-energy balance under the assumption that near the water–air interface the turbulent energy input due to the mean motion is balanced by dissipation.

Now, if we introduce the dimensionless variables

$$(u_i;\ v_i) = (-1)^{i+1} \frac{v_{*i}}{\varkappa} (u_{ni},\ v_{ni});\ z_i = \frac{\varkappa v_{*i}}{f} z_{ni};$$
$$k_i = \frac{\varkappa^2 v_{*i}^2}{f} k_{ni};\ b_i = \frac{v_{*i}^2}{\sqrt{c}} b_{ni}, \tag{19}$$

and then convert in the transformed system from velocities to stresses,* equations (16) and the boundary conditions (18) and (19) can be rewritten as

$$\frac{d^2\eta_{ni}}{dz_{ni}^2} + \frac{\sigma_{ni}}{k_{ni}} = 0; \qquad \frac{d^2\sigma_{ni}}{dz_{ni}^2} - \frac{\eta_{ni}}{k_{ni}} = 0;$$

$$\frac{\eta_{ni}^2 + \sigma_{ni}^2}{k_{ni}} - \mu_{0i} k_{ni} \frac{dS_{ni}}{dz_{ni}} - \frac{b2ni}{k_{ni}} - \beta \frac{d}{dz_{ni}} k_{ni} \frac{db_{ni}}{dz_{ni}} = 0; \tag{20}$$

$$k_{ni} = b_{ni} \left(z_{0ni} + \int\limits_{z_{0ni}}^{z_{ni}} dz_{ni}/\sqrt{b_{ni}} \right);$$
$$\eta_{ni},\ \sigma_{ni},\ b_{ni} \to 0 \quad \text{at} \quad z_{ni} \to \infty; \tag{21}$$

$$\eta_{ni} = \frac{\rho_2}{\rho_1} \frac{v_{*2}^2}{v_{*1}^2} \eta_{n2}; \qquad \sigma_{ni} = \frac{\rho_2}{\rho_1} \frac{v_{*2}^2}{v_{*1}^2} \eta_{n2};$$
$$-\frac{d\eta_{ni}}{dz_{ni}} + \frac{\varkappa V_{gi}}{v_{*1}} = \left(\frac{\rho_1}{\rho_2}\right)^{1/2} \left(\frac{d\eta_{n2}}{dz_{n2}} + \frac{\varkappa V_{g2}}{v_{*2}}\right);$$
$$\frac{d\sigma_{ni}}{dz_{ni}} + \frac{\varkappa U_{gi}}{v_{*1}} = -\left(\frac{\rho_1}{\rho_2}\right)^{1/2} \left(\frac{d\sigma_{n2}}{dz_{n2}} - \frac{\varkappa U_{g2}}{v_{*2}}\right); \tag{22}$$
$$b_{ni} \to 1 \quad \text{at} \quad z_{ni} \to 0.$$

Here $\eta_{ni} = k_{ni} \frac{du_{ni}}{dz_{ni}};\ \sigma_{ni} = k_{ni} \frac{dv_{ni}}{dz_{ni}}$ are the dimensionless frictional stress components; $\mu_{0i} = -\frac{g}{S_i} \times \frac{\varkappa^2 \Pi_{0i}/\rho_i c_i}{f v_{*1}^2}$; Π_{0i} is the turbulent heat exchange

*The advantages of this technique, suggested by Monin [37], are: (1) it eliminates two dimensionless variables, namely $\varkappa V_{gi}/v_{*i}$, and $\varkappa U_{gi}/v_{*i}$, and (2) it converts the boundary conditions (18) from the level z_{0i} to level $z_{ni} = 0$. This change in the lower integration limit is justified by the fact that the frictional stress and turbulent energy are virtually independent of height in the range of z_{ni}.

between the sea surface and the atmosphere (at $i=1$) or the turbulent mass flow on the sea surface (at $i=2$); $\rho_i c_i$ is the volumetric heat capacity; β is a universal constant consisting of α_b and c and equal to 0.54.

Laykhtman has shown that problems (20)-(22) can be reduced to a single-layer problem. The latter has been numerically solved in [1] and [33] for the case of $k_i \frac{dS_i}{dz_i} = \text{const}$ and the case in which dS_i/dz_i is approximated by a given power function of height. In [1] and [33] all the universal functions were obtained by solving Eqs. (20) under the boundary conditions (21), but assuming other conditions [not (22)] hold at the lower boundary of the combined layer. Thus, instead of (22) it was assumed in [1] and [33] that $\eta_n = 1$, $\sigma_n = 0$, and $b_n = 1$ at $z_n \to 0$.

To represent the desired system of universal functions in terms of functions of the single-layer problem, an auxiliary coordinate system $(x_1'y';\ z_i)$, is introduced. In this system, the x' direction and the scales v_{*i} are so selected that at $z_{ni} \to 0$, $\eta_{n1} = \eta_{n2} = 1$ and $\sigma_{n1} = \sigma_{n2} = 0$. In conformance to the first two conditions in (22), this means that

$$\frac{\rho_2}{\rho_1} \frac{v_{*2}^2}{v_{*1}^2} = 1 \tag{23}$$

and that the x' axis of the auxiliary coordinate system points along the shear stress direction and makes an angle with the x axis of the standard coordinate system $(x,\ y,\ z_{ni})$

The scales of v_{*i}, and the angle α between the geostrophic wind direction and shear stress can be derived from the heretofore unutilized conditions of continuity of velocity at the water–air interface [that is, the third and fourth relations in (22)]. The resultant expressions for the geostrophic friction coefficient $\chi = \frac{v_{*1}}{\varkappa_{G_1}}$ and angle α are

$$\chi = (1 - 2n\cos\gamma + n^2)^{1/2} \left[\left(\frac{d\eta_{n1}}{dz_{n1}} + \sqrt{\frac{\rho_1}{\rho_2}} \frac{d\eta_{n2}}{dz_{n2}} \right)^2 + \left(\frac{d\sigma_{n1}}{dz_{n1}} + \sqrt{\frac{\rho_1}{\rho_2}} \frac{d\sigma_{n2}}{dz_{n2}} \right)^2 \right]^{-1/2}_{\substack{z_{n1}=z_{0n1} \\ z_{n2}=z_{0n2}}}; \tag{24}$$

$$\tan\alpha = -\frac{1 - n(N\sin\gamma + \cos\gamma)}{N - n(N\cos\gamma - \sin\gamma)}, \tag{25}$$

where γ is the angle between the geostrophic wind and the current at the top of the sea boundary layer;

$n=\frac{G_2}{G_1}$ $N=\left(\frac{d\sigma_{n1}}{dz_{n1}}+\sqrt{\frac{\rho_1}{\rho_2}}\frac{d\sigma_{n2}}{dz_{n2}}\right)\left(\frac{d\eta_{n1}}{dz_{n1}}+\sqrt{\frac{\rho_1}{\rho_2}}\frac{d\eta_{n2}}{dz_{n2}}\right)^{-1}_{\substack{z_{n1}=z_{0n1}\\ z_{n2}=z_{0n2}}}$. If the water-surface roughness z_{01} is fixed, then $z_{0n1}=(\varkappa^2\chi\mathrm{Ro})^{-1}$, whereas if z_{01} is determined from the Charnok formula $z_{01}=m\frac{v_{*1}^2}{g}$, then $z_{0n1}=m\frac{fG_1}{g}\chi$. Here Ro is the Rossby number, $\mathrm{Ro}=\frac{G_1}{fz_{n1}}$; m is a numerical factor [21] approximately equal to 0.035. For z_{0n2} we have in the two cases $z_{0n2}=\left(\sqrt{\frac{\rho_1}{\rho_2}}\varkappa^2\chi\,\mathrm{Ro}\right)^{-1}$; $z_{0n2}=m\sqrt{\frac{\rho_1}{\rho_2}}\frac{fG_1}{g}\chi$, respectively.

The dependence of χ and α on the dimensionless variable mfG_1/g and stratification parameter μ_{0i} is tabulated in [33]. The value of v_{*1} is calculated from a known value of χ and then v_{*2} and the scales of all the desired functions defined in (19) are deduced from (23). Now it becomes easy to compute the vertical distributions of the wind speed, current, and turbulence coefficient from the following formulas:

$$u_1=G_1\cos\alpha+\chi G_1\frac{d\sigma_{n1}}{dz_{n1}};\quad v_1=G_1\sin\alpha-\chi G_1\frac{d\eta_{n1}}{dz_{n1}};$$
$$u_2=G_2\cos(\alpha+\gamma)-\sqrt{\frac{\rho_1}{\rho_2}}\chi G_1\frac{d\sigma_{n2}}{dz_{n2}};\qquad (26)$$
$$v_2=G_2\sin(\alpha+\gamma)+\sqrt{\frac{\rho_1}{\rho_2}}\chi G_1\frac{d\eta_{n2}}{dz_{n2}};$$
$$k_1=\frac{\varkappa^4\chi^2G_1^2}{f}k_{n1};\qquad k_2=\frac{\rho_1}{\rho_2}\frac{\varkappa^4\chi^2G_1^2}{f}k_{n2}.$$

A check has shown a completely satisfactory agreement between the calculated results and measurements of the characteristics of the atmospheric and sea boundary-layer system. As an example, we present a comparison between the dependence of the drag coefficient C_v on the wind speed, as computed from the Laykhtman model and from observation data (Fig. 5).

The effect of the choice of the numerical constants in the above model was discussed by Radikevich [40], who also calculated the mean multiyear seasonal distributions of wind speed, drift currents, turbulent exchange coefficients, and heat, momentum, and humidity fluxes over the waters of the North Atlantic.

An obvious extension of model [31] was the model of the wind-driven ice drift whereby the floating ice was considered as one of the parts of a combined atmosphere–ice–sea system. A somewhat modified version of this model, originally proposed by Laykhtman [32], was used to determine the ice-drift characteristics,

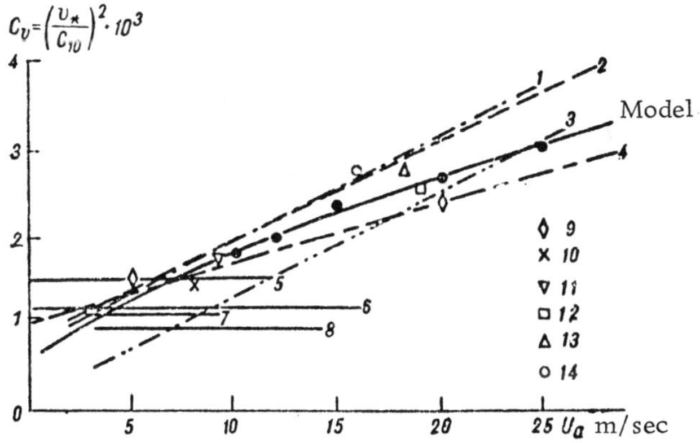

FIGURE 5. THE DRAG COEFFICIENT SHOWN AS A FUNCTION OF WIND VELOCITY, ACCORDING TO LAYKHTMAN [33]:

1–DATA OBTAINED FROM ZUBKOVSKIY AND KRAVCHENKO, 2–SHEPPARD, 3–FRANCIS, 4–DEACON AND WEBB, 5–BROOKS, 6–LACOMB ET AL., 7–FLEAGLE ET AL., 8–DEACON, 9–WILSON, 10–MCELROY, 11–VINOGRADOVA, 12–CALLAGHAN, 13–HELLSTROM, 14–PULMAN AND LAURELL; THE SOLID CURVE REPRESENTS DATA OBTAINED FROM LAYKHTMAN'S MODEL.

including the effect of gradient currents [4]. In brief, this model involves the following. By proper rotation of the coordinate axes, the equations for the air and sea boundary layers are reduced to equations of the single-layer problem, whose solution is known [1, 33]. The solutions are coupled by means of equations for the ice motion and the conditions of matching velocities at the air–ice and ice–water interfaces. As a result, the determination of the ice-drift characteristics reduces to solution of a system of four transcendental equations. Analysis of this system shows that, with a neutral stratification of the air and sea, the velocity and direction of ice drift are functions of three dimensionless variables, namely $fM/\varkappa^2\rho_1 G_1$; G_1/fz_{01} and z_{01}/z_{02}.

Numerical calculations at fixed values of the above variables indicate [4, 32, 33] that the ice-drift coefficient c_0/G_1 and the angle between the drift direction and ground isobar are virtually insensitive to the variations of z_{01}/z_{02}. Of the other variables, $(fM/\varkappa^2\rho_1 G_1)$ defines the velocity and affects but slightly the direction of ice drift, whereas the Rossby number G_1/fz_{01} has a somewhat stronger effect on the deviation of the angle of the ice drift from the isobar.

a)

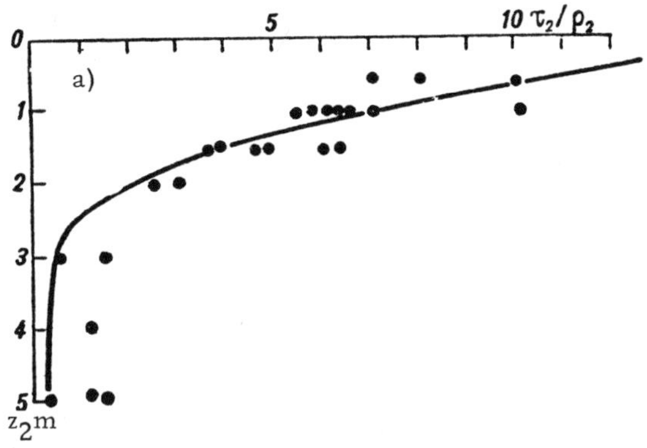

b)

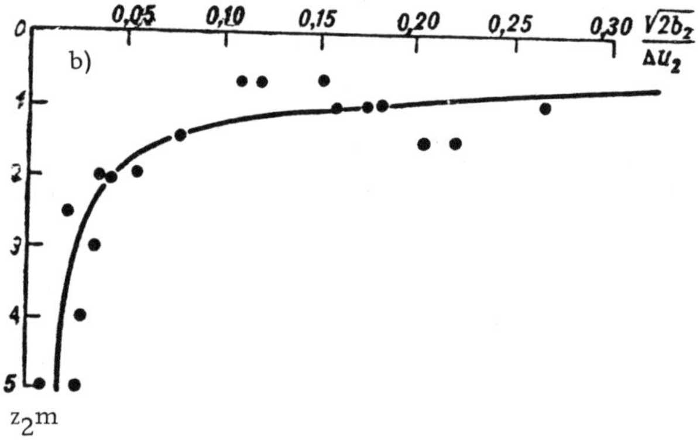

FIGURE 6. VERTICAL PROFILES OF THE SHEAR STRESS (6a) AND KINETIC ENERGY OF TURBULENT FLUCTUATIONS, NORMALIZED TO THE DIFFERENCE OF THE MEAN VELOCITIES (6b) IN THE SUBICE-SEA BOUNDARY LAYER, ACCORDING TO YEGOROV. THE POINTS DENOTE EMPIRICAL DATA FROM KOLESNIKOV, ET AL. [25]; THE SOLID CURVE IS A PLOT OF DATA COMPUTED FROM LAYKHTMAN'S MODEL.

Interesting results have been obtained for the turbulent energy balance of the subice layer of the sea [3]. At a neutral sea stratification, most of the kinetic energy generated by turbulent fluctuations is stored within a 3-m water layer adjoining the lower edge of the ice. Here the generation is almost completely balanced by viscous dissipation. The diffusional transport of turbulent energy within the top sea layer is negligible. This pattern of vertical distribution of the components of the turbulent energy balance is confirmed by Fig. 6 (a, b), which compares the calculated turbulent friction stress τ_2/ρ_2 and the kinetic energy of turbulent fluctuations b_2 with measurements made from the drifting ice station *Severnyy polyus (North Pole).*

V. SOME UNSOLVED PROBLEMS

Our description of models of the system of atmospheric and sea boundary layers would be incomplete if we failed to mention the following.

We noted above that the conditions that are assumed to exist at the interface between the two media are usually not specified on the physical surface of the sea itself but at some roughness level z_{01}, whose magnitude is either fixed or, at best, is determined from the known Charnok formula $z_{01} = m v_{*1}^{2}/g$. As Kitaygorodskiy [21] showed, however, the proportionality factor m in this formula is not a universal constant but a function of the wave state. It is this factor that explains the fairly large differences in the values (ranging from $1.12 \cdot 10^{-2}$ to $7.9 \cdot 10^{-2}$) obtained in working up experimental data by means of the various models. Thus, the first step in developing new physical theories of the interaction between the boundary layers should be definition of the roughness level at which the boundary conditions would be specified.

Further, the boundary conditions themselves should be reviewed. The fact is that the water–air interface is not fixed, but tends to oscillate about its mean value so that at any given time different points on the same horizontal plane that adjoins the mean level may be in air and in water. The farther this horizontal plane is from the mean level of free surface, the sharper the variation of the air-to-water contact area ratio on it. If one then adds to this uncertainty the difference in the statistics of the fluctuations in the two media, then the procedure of averaging quantities in the zone of contact between air and water loses all physical obviousness. The only clear thing here is that the processes occurring in this zone cannot be described by the averaged equations for the air and sea boundary layers.

Because of the above, a new transitional layer, to encompass the fluctuations of the free sea surface, was suggested [21-23, 34, 35]. Given the large gradients of all physical properties within this layer, it cannot be reduced to a single plane while also assuming a continuous transition from one medium to the other. It would appear that the assumption of a discontinuous change in some properties across the air–water interface would be more correct. Unfortunately, this very important problem of the physics of air–sea interaction has not yet been tackled. There are no observational data to give even a very approximate idea about the structure of the transitional layer,* though some very controversial estimates of the momentum flux jump in the transitional layer were reported.

*An attempt to describe the structure of this layer within the framework of the theory of a viscous fluid was made in [18], where the transitional layer was assumed to be a single-phase layer of strongly stratified fluid, to which the equations for the turbulent boundary layer apply.

By definition, the transitional layer is one in which the wind acts directly on the waves. Therefore, if we could evaluate the momentum τ_w, transferred by the wind to gravity waves, then we would have the difference between the frictional stress τ in the air boundary and τ_w, a difference which will describe the desired momentum flux at the bottom of the transitional layer, that is that fraction τ_d of the total momentum flux which is expended on maintaining drift currents. According to the data of various authors (the summary of these data may be found in Kitaygorodskiy's monograph [21]), τ_w ranges from 10 to 60% of τ and, hence, only 40 to 90% of the total momentum flux in the atmospheric boundary layer is expended on generation of drift currents.

The consequences of such an appreciable momentum flux jump in the air–sea contact zone have not yet been theoretically analyzed, even though this is not very difficult, given the existing models, and would be very helpful, among others, in estimating the momentum flux jump that would be optimum for purposes of calculations of drift currents.

In conclusion, let us discuss the controversial problem of the sources of the small-scale turbulence in the sea boundary layer and the associated problem of formulation of the upper boundary condition for turbulent energy.

Most of the above models of air and sea boundary layers assume that the turbulent energy in the sea surface layer is generated only by interaction of Reynolds stresses with mean-velocity shear. For this reason, absence of the diffusive flux (or input) of turbulent energy was made the boundary condition on the free ocean surface. However, there exists another viewpoint, first expressed by Kitaygorodskiy [21], namely that the principal mechanism responsible for the turbulent energy in the sea surface layer is the diffusion of turbulent energy from the transitional layer, in which the breaking of gravity waves converts their energy directly into turbulent energy. This theory was the basis for a novel "diffusional" model of the sea boundary layer [22, 23].

Thus, there exist two alternative approaches to the modeling of processes occurring in the top layer of the sea. It would, of course, be most politic to assume an in-between position and assert right away that the two mechanisms are equally valid. However, we shall make an attempt to understand the reasons for revising the earlier ideas about the mechanism of turbulent energy generation in the surface layer of the sea.

There are two such reasons: (1) the energy E_w that is transferred by wind to the waves is at steady sea state, much greater than energy E_d used to maintain pure drift currents; this defines the ratio E_w/E_d from above, while $E_w/E_d \approx 1$ is characteristic for estimates from below; (2) in the Ekman model of the boundary layer (shear flow), one cannot obtain the dissipation of turbulent energy that is typical of the top layer of the sea [21].

If it is now assumed that at steady sea state the entire energy transferred by the wind to the waves is dissipated as heat, then the natural conclusion is that only the wave-generated turbulence, namely that due to breakup of the sharp-crest gravity waves, can produce the necessary energy dissipation. Hence, the principal mechanism for supply of turbulent energy to the top layer of the sea would be diffusional transport of turbulent energy from the transitional layer to the interior of the sea.

Therefore, to obtain the above conclusion, one needs to keep in mind both of the above reasons. We shall permit ourselves to doubt the second reason, though even the first reason needs verification, as was shown by Byutner [2], who obtained $E_w/E_d \approx 1$ for E_w averaged over the wave energy spectrum.

Thus, let us consider the Ekman boundary-layer model. The pure drift currents in the upper ocean will then be described by the following equations:

$$\frac{d}{dz_2} k_2 \frac{du_2}{dz_2} + f v_2 = 0; \quad \frac{d}{fz_2} k_2 \frac{dv_2}{dz_2} - f u_2 = 0; \tag{27}$$

$$k_2 \frac{du_2}{dz_2} = -\frac{\rho_1}{\rho_2} v_{*1}^2; \quad k_2 \frac{dv_2}{dz_2} = 0 \quad \text{at} \quad z_2 \to 0; \tag{28}$$

$$u_2, \ v_2 \to 0 \quad \text{at} \quad z_2 \to H_2. \tag{29}$$

Here the x axis points along the tangential wind stress.

Let the energy equation be derived from (27) and let us integrate it in terms of z_2 from H_2 to zero. Simple transformations give

$$\left(u_2 k_2 \frac{du_2}{dz_2} + v_2 k_2 \frac{dv_2}{dz_2}\right)_0^{H_2} - \int_0^{H_2} k_2 \left[\left(\frac{du_2}{dz_2}\right)^2 + \left(\frac{dv_2}{dz_2}\right)^2\right] dz_2 = 0$$

or, in accordance with (27) and (28),

$$\frac{\rho_1}{\rho_2} u_0 v_{*1}^2 - \int_0^{H_2} k_2 \left[\left(\frac{du_2}{dz_2}\right)^2 + \left(\frac{dv_2}{dz_2}\right)^2\right] dz_2 = 0. \tag{30}$$

The first term in (30) describes the energy input from wind to drift currents, while the second describes the turbulent energy generation in the friction layer of the sea.

Assume now that there are no vertical density gradients in layer $(0, H_2)$; i.e., buoyancy effects may be neglected. The diffusional flux of turbulent energy at the boundaries of this layer will be assumed to be zero. Then we obtain from the turbulent-energy balance (9)

$$\int_0^{H_2} k_2 \left[\left(\frac{du_2}{dz_2}\right)^2 + \left(\frac{dv_2}{dz_2}\right)^2\right] dz_2 = \langle \varepsilon_2 \rangle H_2. \tag{31}$$

Substituting (31) into (30), we obtain the expression for the mean rate of turbulent-energy dissipation in layer $(0, H_2)$:

$$\langle \varepsilon_2 \rangle = \frac{\rho_1}{\rho_2} \frac{u_0 v_{*1}^2}{H_2}. \tag{32}$$

Just as in [21], we will assume that

$$H_2 = \left(\frac{\rho_1}{\rho_2}\right)^{1/2} \cdot \frac{C_V^{1/2}}{10} \frac{U_a}{f}; \quad v_{*1}^2 = C_U U_a^2,$$

where U_a is the wind velocity measured at a fixed level in the air layer over the water, and C_U is the drag coefficient.

The surface current velocity c_0 will be determined from wind-coefficient data (see Table 1), under the assumption that $c_0 \approx 2 \cdot 10^{-2} U_a$. The other variables ($\rho_1/\rho_2 = 1.3 \cdot 10^{-3}$, $C_U = 1.2 \cdot 10^{-3}$, $U_a = 10 - 15$ m/sec, and $f = 10^{-4}$ sec^{-1}) are from [21]. Then we obtain

$$\langle \varepsilon_2 \rangle = 0.2 \left(\frac{\rho_1}{\rho_2}\right)^{1/2} C_U^{1/2} \cdot U_a^2 \cdot f \approx 2.5 \cdot 10^{-2} \text{ to } 5.5 \cdot 10^{-2} \text{ cm}^2/\text{sec}^3,$$

The order of magnitude of this estimate coincides with the dissipation value of ($\langle \varepsilon_2 \rangle = 10^{-2}$ cm^2/sec^3), which is characteristic of the upper sea layer and was computed from observational data [57]. Even the fact that only 40 to 90% of flux τ is used to maintain the drift currents does not affect the order of magnitude of $\langle \varepsilon_2 \rangle$.

However, not only the mean dissipation, but also the vertical dissipation profile in the sea boundary layer, described by the Ekman model, agree with the empirical data [33] (see Fig. 7). This indicates that the large values of turbulent-energy generation and dissipation, observed in the upper sea, need not be due to the existence of a diffusional flux of turbulent energy from the transitional layer. Going even further than that statement, this flux may be simply assumed to be zero, according to Stewart and Grant [60], who think that the entire small-scale turbulence, generated in the transitional layer by breakup of gravity waves, is completely dissipated *in situ*.

Thus, we have a paradoxical situation, in which two completely opposite models of the sea boundary layer yield the same calculated dissipation of turbulent energy [21] (see Figs. 7 and 8). Whether this is an absurd situation needs to be investigated.

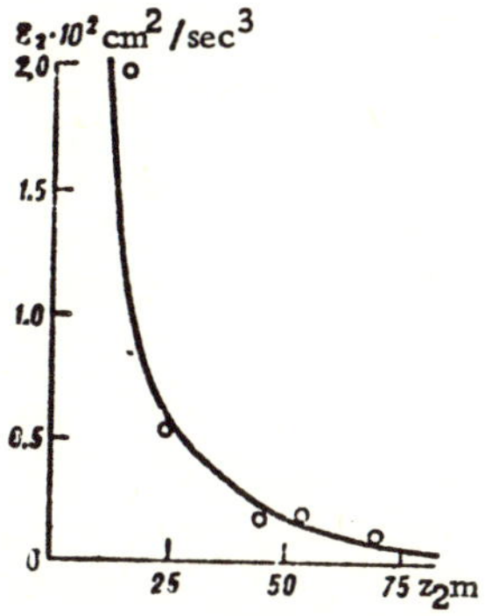

Fig. 7

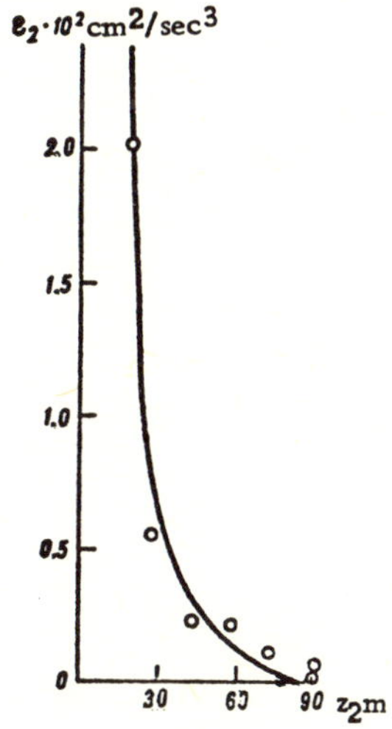

Fig. 8

FIGURE 7. VERTICAL DISTRIBUTION OF THE MEAN RATE OF THE TURBULENCE ENERGY DISSIPATION IN THE TOP LAYER OF THE SEA ACCORDING TO LAYKHTMAN [33]. THE POINTS ARE ON EMPIRICA DATA OF GRANT, MOILLIET, AND VOGEL [57]; THE SOLID CURVE SHOWS THE RESULTS OF CALCULATIONS USING THE LAYKHTMAN. MODEL.

FIGURE 8. VERTICAL DISTRIBUTION OF THE MEAN RATE OF TURBULENT ENERGY DISSIPATION IN THE TOP SEA LAYER, ACCORDING TO KITAYGORODSKIY [21]. THE SOLID CURVE IS A PLOT OF RESULTS CALCULATED FROM THE KITAYGORODSKIY-MIROPOLSKIY MODEL; FOR OTHER NOTATIONS, SEE FIG. 7.

LITERATURE CITED

The number in parentheses refers to the number of the citation in the Soviet Abstracts Journal (Referativnyy Zhurnal) Geophysics 1970.

1. Bobyleva, I. M., S. S. Zilitinkevich, and D. L. Laykhtman, Turbulence in a thermally stratified atmospheric boundary layer. In Atmosfernaya turbulentnost' i rasprostraneniye radiovoln (Atmospheric Turbulence and Radiowave Propagation), Nauka Press, Moscow, 1967, pp. 179-190 (1968, 7B280).

2. Byutner, E.K., Calculation of wind energy used to induce drift currents and waves at various states of the sea surface. Trudy Glav. geof. observ., 1970, No. 226, pp. 25-31.
3. Yegorov, K.L., Effect of background variables on the wind-driven ice drift and the characteristics of turbulence in the subice sea layer. Trudy Leningrad. Gidrometeorol. inst. 1970, No. 40, pp. 93-100.
4. Yegorov, K.L., Floating-ice drift in the presence of gradient current. *Ibid.*, 1970, No. 41, pp. 104-111 (1971, 1B51).
5. Zemtsov, G.S., and B.A. Kagan, Calculation of the vertical velocity component of the pure drift current in the sea friction layer. *Ibid.*, 1967, No. 24, pp. 143-147 (1968, 6B91).
6. Zilitinkevich, S.S., D.L. Laykhtman, and A.S. Monin, Dynamics of the atmospheric boundary layer. Izv. Akad. Nauk SSSR, ser. fiz. atm. okean., 3, No. **3**, pp. 297-333, 1967.
7. Kagan, B.A., Calculation of the surface drift currents. Trudy Leningrad. Gidrometeorol. inst., 1961, No. 10, pp. 98-106.
8. Kagan, B.A., Steady-state drift currents in deep sea. Okeanologiya, **2**, No. 6, pp. 974-980, 1962.
9. Kagan, B.A., Calculation of friction depth in the sea. Trudy Glav. geofiz. observ., 1962, No. 127, pp. 93-98.
10. Kagan, B.A., Thermodynamic interaction in the sea–ice–atmosphere system. *Ibid.*, 1963, No. 144, pp. 183-195 (1964, 12B286).
11. Kagan, B.A., Effect of distribution of turbulence coefficients in the atmosphere and the sea on the velocity profiles of drift currents and wind over the sea. Trudy Leningrad. Gidrometeorol. inst., 1964, No. 17, pp. 128-141.
12. Kagan, B.A., Estimate of turbulent energy dissipation in the pure drift sea currents. Okeanologiya, **4**, No. 1, pp. 3-8, 1964.
13. Kagan, B.A., Theory and calculation of thermodynamic interaction between the sea, ice, and atmosphere. Trudy Leningrad. Gidrometeorol. inst., 1965, No. 20, pp. 105-119.
14. Kagan, B.A., Calculation of water temperature and currents in the North Atlantic by means of the quasistationary model of interaction of the atmospheric and sea boundary layers. Izv. Akad. Nauk SSSR, ser. fiz. atm. okean., **1**, No. 8, pp. 845-852, 1965.
15. Kagan, B.A., Interaction in the North Atlantic of air and sea boundary layers. Trudy Leningrad. Gidrometeorol. inst., 1967, No. 24, pp. 3-21 (1968, 6B61).
16. Kagan, B.A., Response of drift currents and turbulence in the top sea layer to the change in the tangential wind stress. Izv. Akad. Nauk SSSR, ser. fiz. atm. okean., **4**, No. 9, pp. 1004-1007, 1968 (1969, 2B43).

17. Kagan, B.A., and Z.M. Utina, Contribution to theory of thermodynamic interaction between the sea and the atmosphere. Okeanologiya, **3**, No. 2, pp. 150-259, 1963.
18. Kalatskiy, V.I. Effect of the parameters of the transitional layer on the sea–air interaction. Trudy Gidrometeorol. Tsentra SSSR, 1969, No. 51, pp. 43-53 (1970, 2B40).
19. Kitaygorodskiy, S.A., Small-scale interaction between atmosphere and sea (brief review of Soviet research). Okeanologiya, **7**, No. 5, pp. 774-785, 1967 (1968, 5B39).
20. Kitaygorodskiy, S.A., Small-scale interaction between air and sea. Izv. Akad. Nauk SSSR, ser. fiz. atm., okean., **5**, No. 11, pp. 1114-1131, 1969 (1970, 3B247).
21. Kitaygorodskiy, S.A., Fizika vzaimodeystriya atmosfery i okeana (Physics of Air–Sea Interaction) Gidrometeoizdat Press, Leningrad, 1970 (1970, 10B49K).
22. Kitaygorodskiy, S.A., and Yu.Z. Miropolskiy, Contribution to theory of turbulent exchange in the (top) sea boundary layer. Izv. Akad. Nauk SSSR, ser. fiz. atm. okean., 3, No. 11, pp. 1196-1209, 1967 (1968, 6B66).
23. Kitaygorodskiy, S.A., and Yu.Z. Miropolskiy, Turbulent energy dissipation in the surface layer of the sea. *Ibid.*, 4, No. 6, pp. 647-659, 1968 (1968, 12B65).
24. Kolesnikov, A.G., Some results of direct determination of the rated vertical exchange in the sea. In: Nekotoryye problemy i Rezul'taty okeanologicheskikh issledovaniy (Some Problems and Results of Oceanological Research), Press of Acad. Sci. USSR, 1959, pp. 20-28.
25. Kolesnikov, A.G., N.A. Panteleyev, and V.N. Ivanov, Observation of the turbulent entrainment layer under a drifting ice floe. Izv. Akad. Nauk SSSR, ser. fiz. atm. okean., **1**, No. 12, pp. 1310-1318, 1965.
26. Koshmider, G., Dinamicheskaya meteorologiya (Dynamic Meteorology), GONTI Press, Moscow, 1938.
27. Laykhtman, D.L., Patterns of physical processes in the atmospheric boundary layer in the Arctic. In: Sovremennye problemy meteorologii prizemnogo sloya vozdukha (Problems of Meteorology of the Ground air Layer), Gidrometeoizdat Press, 1958, pp. 26-42.
28. Laykhtman, D.L., Wind-induced drift of ice fields. Trudy Leningrad Gidrometeorol. inst., 1958, No. 7, pp. 129-137.
29. Laykhtman, D.L., Formulation of the problem of steady-state structure of the atmospheric boundary layer. Trudy Glav. geofiz. observ., 1960, No. 94, pp. 3-7.
30. Laykhtman, D.L., Fizika pogranichnogo sloya atmosfery (Physics of the Atmospheric Boundary Layer), Gidrometeoizdat Press, 1961.

31. Laykhtman, D. L., Dynamics of the air and sea boundary layers, allowing for interaction and nonlinear effects. Izv. Akad. Nauk SSSR, ser. fiz. atm. okean., 2, No. 10, pp. 1017-1025, 1966 (1967, 3B174).
32. Laykhtman, D. L., Nonlinear theory of the wind-induced drift of ice. *Ibid.*, 4, No. 11, pp. 1220-1223, 1968.
33. Laykhtman, D. L., Fizika pogranichnogo sloya atmosfery (Physics of the Atmospheric Boundary Layer), 2nd edition. Gidrometeoizdat Press, Leningrad, 1970 (1970, 11B52K).
34. Laykhtman, D. L., B. A. Kagan, and R. S. Bortkovskiy, Analysis of hydrodynamics equations describing the sea–atmosphere interaction. Proc. of All-Union Conference on Maritime Meteorology, Gidrometeoizdat Press, Leningrad, 1966 (1967, 9B60).
35. Laykhtman, D. L., B. A. Kagan, and V. V. Timonov, The modes of analysis of air–sea interaction. Meteorolog. issledovaniya, 1968, No. 16, pp. 287-300 (1969, 6B233).
36. Lifshits, N. M., A simple model of unsteady-state interaction between the air and sea boundary layers. Abstracts of papers presented at a conference of young scientists of the Moscow State University devoted to the problems of the ocean. Moscow State University Press, 1969, p. 155
37. Monin, A. S., Dynamic turbulence in the atmosphere. Izv. Akad. Nauk SSSR, ser. geograf. and geofiz., **14**, No. 3, pp. 232-254, 1950.
38. Monin, A. S., Basic consequences of the air–sea interaction. Izv. Akad. Nauk SSSR, ser. fiz. atm. okean., 5, No. 11, pp. 1102-1113, 1969 (1970, 3B246).
39. Panteleyev, N. A., Investigations of turbulence in the surface water of the Antarctic sector of the Indian and Pacific Oceans. Trudy Okeanograf. Komissii Akad. Nauk SSSR, **10**, No. 1, pp. 137-140, 1960.
40. Radikevich, V. M., Analysis of some characteristics of the air and sea boundary layer interaction on the basis of a new theoretical model. Trudy Leningrad Gidrometeorol. inst., 1968, No. 32, pp. 3-15.
41. Stewart, R., Small-scale interaction between the atmosphere and the sea. In: Basic problems of oceanolgoy (Papers presented at the Plenary Session of the 2nd Int. Oceanograph. Congress, Moscow 1966), Nauka Press, Moscow, 1968 (1969, 4B67).
42. Shvets, M. E., Contributions to the theory of sea currents. Meteorol. and Gidrologiya, 1939, No. 3, pp. 3-7.
43. Åkerblom, F., Recherches sur les courants les plus bas de l'atmosphere au-dessus du Paris. Nova Acta Reg. Soc. Scand., ser. 4, **2**, No. 2, pp. 1-45, Uppsala, 1908.

44. Benton, G.S., et al., Interaction between the atmosphere and the oceans. Bull. Amer. Meteorol. Soc., **44**, No. 1, pp. 4-17, 1963 (1963, 10B331).
45. Ekman, V.W., Om jordrotationens inverkan på vindströmmar i havet. Nyt. Mag. für Naturvid., **40**, pp. 37-63, 1902.
46. Ekman, V.W., On the influence of the earth's rotation on ocean currents. Arkiv. Mat., Astron., Fysik, Uppsala—Stockholm, **2**, No. 11, pp. 11-52, 1905.
47. Exner, F.M., Zur Kenntniss der intersten Winde über Land und Wasser, usw. Ann. Hydr. und Marit. Met., **5**, pp. 226-239, 1912.
48. Gonella, J., Interaction ocean-atmosphere. Rapp. nat. trav. franc., 1963-1966, Paris, 1967, pp. 266-269 (1968, 3B45).
49. Gongwer, C., and E. Finkle, Turbulence meter an instrument for oceanographic research. Proc. Instrum. Soc. Amer., **15**, No. 23, pp. 88-93, 1960.
50. Hesselberg, Th., The Ekman spirals. Arch. für Meteorol. Geophys. und Bioklim., **A7**, pp. 329-343, 1954.
51. Hughes, P.A., A determination of the relation between wind and sea-surface drift. Quart. J. Roy. Meteorol. Soc., 82, No. 354, pp. 499-502, 1956.
52. Kraus, E.B., Wind stress along the sea surface. Advances in Geophysics, **12**, pp. 213-255, 1967 (1968, 6B60).
53. Kraus, E.B., Sea—air interaction, Trans. Amer. Geophys. Union, **48**, No. 2, pp. 581-584, 1967 (1968, 1B55).
54. Neumann, G., and W.J. Pirson, Principles of Physical Oceanography, Prentice-Hall, 1966.
55. Pandolfo, J.P., A numerical model of the atmosphere-ocean planetary boundary layer. Proc. WMO/JUGG Sympos. Numerical Weather Prediction, Tokyo, 1969, pp. 1131-1140 [1970, 2B39].
56. Pandolfo, J.P., Motions with inertial and diurnal period in a numerical model of the Navifacial boundary layer. J. Marine Res., **27**, No. 3, pp. 301-317, 1969 (1970, 10B56).
57. Phillips, O.M., Dynamics of the Upper Ocean. Cambridge Univ. Press, 1966.
58. Roll, H.U., Physics of the Marine Atmosphere. Academic Press, N.Y.—London, 1965.
59. Schmidt, W., Der Massenaustausch in freier Luft und verwandte Erscheinungen. Verlag Henrini Grand, Hamburg, 1925.
60. Stewart, R.B., and H.L. Grant, Determination of the rate of dissipation of turbulent energy near the sea surface in the presence of waves. J. Geophys. Res., **67**, No. 8, pp. 3177-3180, 1962.
61. Tomczak, G., Neuere Untersuchungen mit Treibkörpen zur Bestimmung des Windeinflusses auf Oberflächenstromungen

im Meer. Ber. dtsch. Wetterdienstes, **91**, No. 2, pp. 18-25, 1963.

62. Watson, A. G. D. Air–sea interaction. Sci. Progress, **56**, No. 223, pp. 303-323, 1960 (1969, 4B59).
63. Yoshihara, H., Sea–air interaction: a simplified model. J. Atmosph. Sci., **25**, No. 5, pp. 729-735, 1968 (1969, 4B289).

SEA STRAITS

D. M. Tolmazin

The beginnings of the theory of sea straits go back to S. O. Makarov, who, on the basis of relatively skimpy data on hydrology of straits, suggested in 1882 that the easiest theory of sea currents to develop would be that for a two-dimensional channel, because such a theory could be tested in one of the straits where both drift and gradient effects are present. Makarov's prediction proved to be right to some extent, in that even such specifically "three-dimensional" current theories as P. S. Lineykin's the baroclinic model of the ocean, the method of mean sections of Shtokman [39] , as well as his model of flow (which takes account of baroclinity by means of an "influence function"), and Fel'zelbaum's theory of currents in seas of low and average depths all started as two-dimensional models. The only reason why the results computed from these theories could not be compared with the actual flow of water straits is the inherent limitation imposed by key basic assumptions of such models (such as infinite depth of the strait).

In this century the sea straits have been fairly well investigated. While significant contributions were made by S. O. Makarov, A. Mertz, L. Moller, A. Defant and others, the most complete generalizations of the hydrology of straits was formulated by N. N. Zubov in his *Osnovy ucheniya o prolivakh Mirovogo okeana* (The Foundations of the Theory of Ocean Straits), Geografgiz Press Moscow, 1956. This book presents a classification of straits in terms of their features, describes processes observed in straits, and shows how straits affect the density distribution in the sea.

The present review is based on Zubov's classification of straits in terms of water exchange. We shall describe three basic groups, to wit: a) straits with vertical water exchange; b) straits with horizontal water exchange; and c) shallow straits. Each group exhibits a specific hydrology, which in turn engenders a specific set of practical and theoretical problems. A summary of these problems precedes the account of the results of scientific studies carried out during 1967 to 1970.

In recent years the interest in the study of straits has increased in connection with intensified economic activity of man. For example, irretrievable loss of river runoff diverted away from inland seas, the deepening of ship channels, and intensification of fishing, among others, have led to the necessity of studying water and salt exchange through straits and in some cases to estimate its changes following the creation of artificial narrows. Some theoretical analysis of current dynamics and turbulence in straits is involved in these practical problems of rebuilding the natural environment. Several complex regional studies of currents and of the thermohaline structure of water were undertaken in order to define the water exchange between mediterranean and inland seas and to estimate the effect of large islands on the circulation of oceanic currents. The idea of regulating the climate in individual areas, or of the Earth as a whole, is based on water exchange through large straits. So far these tentative ideas have not gone beyond popular-science hypotheses, but they are of some interest because they show that it is theoretically possible to influence the climate of the Earth.

STRAITS WITH VERTICAL WATER EXCHANGE

The Bosporus, the Dardanelles, Gibraltar, the Bab el Mandeb, and Sunda Straits, connecting seas of different water and heat budget, may be assigned to this group. Vertical water exchange is also characteristic of some fairly deep estuaries and many fjords of Alaska and Norway.

The principal feature of the water circulation in this type of strait is the presence of a two-layer system of currents, caused by the fact that the pressure gradient changes sign with depth. Near the surface this gradient is controlled by the longitudinal level gradient, whereas at depths it is induced by the density difference in the water bodies linked by the strait. The significant stratification in such straits is due to the superposition of water masses of different origin, and so they are called stratified straits.

Most theoretical and interdisciplinary hydrological work on this type of strait deals with current fields, water exchange through the strait and its effect upon the thermohaline structure and gas content of the water in the inland sea and the adjacent ocean. The relative simplicity of the two-dimensional flow pattern in the strait makes possible experiments to determine some of the turbulence parameters which cannot be deduced by direct measurements.

The earlier studies of stratified straits employed a two-layer model of water flow developed by Defant (A. Defant. Physical Oceanography, Vol. I. Permagon Press, 1961) and subsequently modified by the present author. Filippov [35] has shown that this model describes quite well the vertical velocity profiles and the position of the interface between currents actually observed in the Bosporus. A similar model can be used for analysis of seasonal flow in some straits in the Far East [19]. However, the model has proved to be totally useless in cases in which currents and countercurrents occur in the same horiziontal plane.

Some schemes for calculation of velocities of deep currents in stratified straits and their application to analysis of the Lower Bosporus current are described in [35] and are of considerable interest. These calculations indirectly take into account the effect of baroclinity. It is shown that density currents can be compared with suspension (turbidity) currents whose theory is known and which have been studied by observation. However, it is difficult to find empirical coefficients of friction for the relatively small density differences between waters of different origin that interact within straits. The hydraulic approach to analysis of deep-deep currents in stratified straits is made difficult by another fact, namely that such straits lack a clearly defined overall slope, whereas, as will be shown below, the theory of hydrology of inland basins is based on the assumption of flow of density currents around sills in or at the exits of straits.

Because of the above, better results, in terms of understanding of current dynamics and field densities, would be expected from hydrodynamic models of currents in straits. Tolmazin [69] discusses one of these models, described in detail in [32]. He solves the two-dimensional equations of flow and turbulent diffusion for a strait of rectangular cross section, in which the width is constant and the depth variable. The heights of levels are specified at the strait's ends. The solved-for functions of the problem are coupled with one auxiliary Q function, for which a nonlinear equation of the fifth order, in terms of the vertical coordinate z, has been obtained. A tangential wind stress at the surface, as well as adhesion at the sea floor are specified, as are two density conditions. The fifth condition is obtained from the equation of continuity. In its dimensionless form, the equation contains a small parameter at the highest derivative, which indicates the existence of boundary layers in the problem. Tolmazin derived a self-similar solution of the Q equation, based on a characteristic specification of longitudinal changes of bottom topography and wind in different straits. The calculated velocities, vertical-velocity profiles, and boundaries of currents and water masses are in fair agreement with measurements in the Bosporus, Gibraltar, and Bab el Mandeb. The three-layer current structure, observed in some periods of the year in the Bab el Mandeb Strait, was also successfully modeled.

Rattry [59] suggested a somewhat different baroclinic flow model for a stratified canal, applicable to some fjords and straits of Alaska and Canada, the Georgia and Juan de Fuca Straits, the Alberny and Admiralty Fjords, among others. Rattry showed that the temperature and salinity perturbations decay fairly rapidly with depth, as does the vertical-velocity gradient. This led him to assume that the deviations of the basic characteristics from constancy disappear at very great depths. The hydrodynamic problem is solved for two velocity components (horizontal and vertical) and for salinity deviations. Flows with different vertical directions are generated by river flow at the surface and gradient-driven countercurrents produced by longitudinal density gradients. The problem reduces to the solution of a system of two nonlinear equations for two generalized functions related to velocity and salinity components. Self-similar transformation of the equations reduces them to regular ones, with constraints imposed on the effect of longitudinal wind stress on the friction and diffusion coefficients. The regular equations were solved by expansion into a power series with a factor exponentially decreasing with depth. A comparison of calculated velocity and salinity profiles with observations shows a good agreement in the case of several fjords and straits.

The above-described problems lead to more realistic models, which better describe the mechanism of current generation. It appears that the problems of water flow in straits would be formulated more completely if several controlling factors, such as level differences or horizontal density gradients, were to be calculated from data on large-scale processes occurring in the atmosphere and in basins joined by such straits.

Interdisciplinary hydrologic and hydrochemical investigations of individual stratified straits show a general tendency—namely attempts to define the quantitative or qualitative characteristics of water exchange, based on current, temperature, salinity, and oxygen measurements, and then to trace the subsequent "fate" of abyssal waters and their influence upon the density structure of seas adjacent to the given strait. Whether narrower and more specific problems such as the estimation of the effect of morphometry on the dynamics of abyssal currents, the determination of the microstructure of currents, or calculation of turbulence characteristics can be solved depends strongly on how much is known about the water dynamics in an individual strait and on what specifically a given study is trying to achieve. Because of this, studies involving specific observations are best summarized strait by strait.

Bosporus is one of the best understood straits, primarily because of Bogdanova's monograph [3] on the effect of Bosporus on the hydrology and biology of the Black Sea. Bogdanova's review of the hydrometeorology of the Bosporus and the pre-Bosporus* area is the most detailed one in existence. It summarizes the entire

*That is, the area of the Black Sea adjacent to the Bosporus.

discussion on the role of the northern sill of the Bosporus and the fate of Mediterranean water in the Black Sea. Bogdanova has shown conclusively [43] that the flow of Mediterranean water in the Black Sea has been observed under most unfavorable conditions and that it takes a fairly complicated path upon its issue from the strait. Thus it first flows northwestward and is then entrained by the general circulation to the northeast. To clarify seasonal changes of water exchange, Bogdanova reviewed all the available observations by Mertz and others and analyzed the nature of changes of level gradients between seas. She found that the reason for the raise of the current boundary surface above the sill is the horizontal divergence in the pre-Bosporus area, which is due to bifurcation of the nearshore flow on approach to Bosporus.

In connection with Pektas' statement that Mediterranean water contributes to the isolation of the abyssal water of the Black Sea as they spread an intermediate layer at a depth of 150 to 200 m, Bogdanova noted that the specific properties of the intermediate layer, namely, the increase of the salinity gradient beneath the cold parting layer, the temperature increase below the depth of convection depth, and the presence of oxygen below that depth, are not caused by Mediterranean water but by inflow of cold water that has been cooled in winter in the northwestern Black Sea. The Mediterranean water, which can be traced in terms of its temperatures (9.92° to 10.14°C), salinity (22.12 to 22.30 ‰) and oxygen content (0.11 to 0.37 ml/liter) enters the Black Sea near the Bosporus at a depth of 500 to 530 m and forms the abyssal water mass [9]. The study of this process is still in progress [22, 33].

In a chapter of Bogdanova's book dealing with the hydrochemical indicators of Bosporian influence [12], Dobrzhanskaya states that Bosporus contributes merely 0.5% of the average annual oxygen supply to the upper 150-m layer of the Black Sea, but that the boundary of hydrogen sulfide zone drops down to a depth of 350 to 400 m in this area. The effect of Bosporus water on the concentration of biogenic materials, particularly phosphates, is virtually insignificant, especially in areas which are at some distance from the strait.

The study of the influence of Bosporus on the Black Sea is a matter of major importance not only for better understanding of the structure of the strait water and biology, but also for forecasting the future hydrology of the Black Sea, since one of the components of the fresh-water budget of that sea—the river runoff—will be substantially reduced in the coming decades. This problem is dealt with in [10], where the future salinity of the Black Sea is predicted from an unsteady-state equation of the salinity budget (not derived in the paper). Assuming an annual runoff loss of 150 to 170 km^3, the salinity will tend to 33 ‰. However, it will take as much as 7000 years to reach this level. The result will be that the Black Sea would come to resemble the Mediterranean, for its depth of convection would rise, as will its average temperature. The hydrogen sulfide layer should subside, and even the color of the sea may change toward the blue.

The two-dimensional structure of the property fields in the Bosporus allows one to calculate the parameters of the vertical turbulence with the aid of simplified dynamics and diffusion equations. Jakobsen, in 1913, was the first to perform such calculations for the Kattegat. Bogdanova and Tolmazin calculated the eddy diffusion coefficients K_z for salts by a steady-state, finite difference diffusion equation, which takes into account the vertical salt advection. The vertical velocity is determined from a simplified continuity equation, the calculations being based on observations of Mertz. The minimal coefficients K_z, equal to units in CGS system, were obtained for the maximum stability layer, whereas above and below this layer K_z is one to two orders larger.

Bogdanova et al. used the same technique in [4] to calculate conditions in the pre-Bosporus area; however, in this case the horizontal diffusion of salts was taken into account. The so-obtained K_z values formed the basis of salt-balance calculations for the wedge of lower Bosporus water. This study produced the criteria for estimating the maximum range of penetration of high-salinity water into the Black Sea.

Significant studies on the deep current flowing from the Gibraltar into the Mediterranean sea were recently completed. The most important of these was the so-called Mediterranean Discharge project started in September–October 1965 under the auspices of the NATO oceanographic subcommittee. Two lengthy studies [53, 50] summarize the results of observations of hydrological variables by the R/V *Calypso* and R/V *Origny*, with seven other ships from Norway, Portugal, Belgium, and France participating in this study. The main task was to define the quantity and characteristics of Mediterranean water flowing into the Atlantic. Plots of some characteristics, reflecting the behavior of Mediterranean water into the Atlantic away from Gibraltar are given in [50, 53].

Later observations are generated in [55, 49]. Thus, paper [55] presents a comparative analysis of 1967 observations and data of earlier years. It was found that the discharge from the Mediterranean is about $170 \cdot 10^6$ m^3/sec. The chart showing the distribution of temperature, salinity, and currents indicates the Mediterranean water flows along sea-floor depressions, the main, northwestward branching into streams flowing into submarine trenches.

Paper [49] describes the currents within the stream of Mediterranean water with the aim of defining the distribution of alluvium. The main stream separates from the sea floor 250 km away from the strait. From that point it can be traced, because of its salinity and temperature, at depths of the order of 1200 m. Because of the high velocity of the deep current, the sea floor in the Strait of Gibraltar is free of sediments, but sand is deposited as wavelike dunes 5 m high and 100 m long, at a distance of 150 km from the strait. The frequent ruptures of cables west of Gibraltar are due to shifts of bottom deposits.

The Strait of Gibraltar is fairly wide, and so because of the presence of geostrophic effects the isopycnic surfaces in it may exhibit transverse gradients. This allows one to use Margules' formulas [70] to analyze the relations between the velocities of the top and lower currents. The 14° isotherm is taken as the boundary surface, and the gradient of the sea surface is calculated from data obtained at observation posts. Calculations of the average velocities and flows of the top and lower currents are also given in [49].

In the recent decade, the Bab el Mandeb Strait has been subjected to a detailed study. A wide range of hydrographic studies was carried out in 1962-1963 by the Institute of Biology of Southern Seas of the Ukrainian Academy of Sciences using the R/V *Akademik Kovalevskiy*, and in 1964-1965 by the Institute of Marine Studies of University of Kiel using the R/V *Meteor*, in the Red Sea and the Gulf of Aden. In various seasons of 1961-1966, ships of the Azov and Black Seas Research Institute of Sea Fisheries and Oceanography also undertook extensive observations in the above areas. The results of these studies are summarized in [2, 29, 30, 38, 40, 62, and 63], of which the theoretically most important are the papers by Siedler [63] and Khimitza [38].

Briefly these expeditions found that the currents and water exchange in the Bab el Mandeb are a function of the wind and evaporation in the Red Sea, and change in the course of the year. There is a fairly sharp relationship between current structure and other hydrological variables on the one hand, and the monsoon circulation on the other. During the winter and spring the evaporation in the Red Sea is relatively slow and the winds are southerly. One then observes a two-layer current structure with a horizontal boundary between flows. However, a current from the Gulf of Aden into the Red Sea may appear in the lower layer, and then the boundary between the currents flowing in the various directions may become almost vertical. Khimitza [38] calls this a mixed structure and ascribes the horizontal branching of flows to the deflecting force of the Earth's rotation. In the summer, when the evaporation rate is low and the winds north to northwesterly, the two main currents described above become supplemented by a drift current at the surface, and thus there exists a three-layer current system. Most of the authors [30, 38, 40] note that during this period the middle current fills almost the entire cross section of the strait, while the saline and warm Red Sea water flows into the Gulf of Aden along the strait bottom. Its future fate depends on the circulation existing within that Gulf. In winter, this water penetrates into the intermediate layers of the Indian Ocean, but becomes freshened and cooled along its path. Siedler [63] was able to elucidate the rather complex thermal microstructure of the water at the strait exit, and concluded that the temperature fluctuations of several degrees and the salinity fluctuations of the order of 0.1‰, observed in the analysis of time series of data, are due to internal waves whose periods approach those of tidal waves.

Among other straits of the same type, Kattegat, Skagerrak, the Straits of Sunda, Little Belt, and Great Belt have been extensively explored [54, 66-68].

Water exchange in the vertical plane can occur in many other straits, if the density and level differences in the adjacent basins are due either to excess flow or wind-driven tide. There is evidence of water exchange between the Adriatic and Ionian Seas [71].

We need to mention two more papers, dealing with long-term special observations in fjords and with calculations of currents and turbulence in them. Thus, Gade [48] describes the temperature fields and variability of salinity in the Oslofjord over a multi-year period. The coefficient of vertical eddy diffusion was estimated with the aid of the phase-shift method.

Sternberg [64] reports on the extensive observations on near-bottom flows, carried out in the above-cited [59] straits and fjords of Alaska (Juan de Fuca, etc.; six straits in all). Current observations at several horizons were supplemented by topographic observations of the sea floor by means of underwater television cameras and stereophotography, so as to measure the roughness and structure of the floor. A logarithmic velocity profile was observed in all basins, but is most pronounced in the zone of strong currents. Long-term observations of current profiles yielded data for calculation of Reynolds numbers under different sea-floor morphologies. Sternberg classified currents into hydrodynamically unstable at small Re, and hydrodynamically rough at large Re, and has shown that larger sea-floor irregularities affect the characteristics of the near-bottom turbulence.

In summarizing this section, we need to say that most of the work so far done on straits with water exchange induced by density gradients has dealt with steady-state and transitional fields.

Transient phenomena in straits, such as internal waves and induced current fluctuations due to variability of wind circulation have not been adequately studied, even though the importance of transient effects in wide straits had been emphasized in the earlier studies of Defant, Schott, Vercelli, and Lacombe.

STRAITS WITH HORIZONTAL WATER EXCHANGE

It was shown a long time ago that in some fairly wide straits such as the Davies, Bering, Formosa, and Denmark there exists a characteristic distribution of currents and temperature, in that the current velocity and temperature is always higher on one coast than on the other. Countercurrents separated from the main current by an almost vertical boundary face had also been observed. In his study of the Strait of Formosa, S. O. Makarov showed that a horizontal stratification of temperature and salinity may be so distinct that the isotherms, corrected for tide, could serve as a kind of fairway guide for ships. The first attempt to analyze this phenomenon was made by Shtokman. His strait model of the Bering Strait* simulated the real currents and countercurrents existing at a transversely inhomogeneous wind field. He also attempted to explain qualitatively the anomalous currents about islands such as Iceland and Madagascar, which are situated at the margins of the oceanic circulation systems [39]. Shtokman's intuitive approach does not negate the importance of nonlinear effects (for instance, longitudinal advection of momentum) in currents around islands whose scale is of the same order of the transverse dimension of the flow. To check this theory, Shtokman calculated the steady state of currents in a multiply connected rectangular area with the aid of Felzenbaum's model [22], assuming that flow is induced by wind with a linear transversal shift. Shtokman found that the presence of one or more islands does not generate overall circulation anomalies in the area between the mainland and the island. But, the surface-currents field exhibits closed lines to the left of the island (or islands) only under some special circumstances, thus partially confirming the theory. Shtokman [22] found that a more nonlinear than assumed decrease of wind speed, localized in a narrow zone, is the factor responsbile for stable anomalies. However, in evaluating conclusions derived on the basis of Felzenbaum's model, one must allow for possible errors due to substitution of the general

---

* V.B. Shtokman. The wind's influence on currents in the Bering Strait; causes of their high velocities and prevailing northern direction. Trudy Inst. okeanolog. Akad. Nauk SSSR, 1957, p. 25.

formula for velocity by a simpler one that neglects the Coriolis force. Application of Felzenbaum's model to a topographically complicated area [33] has shown that neglect of the Coriolis force at fairly deep sites has a pronounced effect on the velocity field.

Transverse inhomogeneity does not appear to be the only cause responsible for the above anomalies. The boundary layer also exerts an important effect. An island near the mainland, situated at the margin of a large ocean gyre whose branch must be treated as a "potential current" must have an effect decreasing the width of the displacement near the mainland. These qualitative analyses can be checked out on a theoretical model, examining separately the boundary layer and the inner area, with a subsequent matching of the solutions (Carrier's approach). Under these circumstances, an island is in the inner area where the nonlinear boundary effects may be neglected. However, in this approach Felzenbaum's model will be inadequate, in that apart from wind-generated gyres one must take into account the diffusion of the relative vortex (lateral friction), and then the effect of the island will be that of a solid rather than a liquid boundary.

Whereas the above-described theoretical analyses of the straits did not aim at creating a technique for computing the currents within such straits, but merely attempted to explain the nature of bilateral currents, the comprehensive regional studies undertaken there tried to describe the currents and water transport quantitatively, and to test the applicability of the dynamic method for current calculations. Again, for convenience, we shall describe the various studies strait by strait rather than in terms of individual topics.

The first fundamental studies of currents were carried out in the Denmark Strait [41, 42, 65, 70]. It was found that the velocity of the Greenland current in the strait decreases from 14 to 4 cm/sec with distance from the shore and that the current veers to the east [41]. At greater depths during the summer, baroclinic effects are fairly significant [42]. The abyssal currents were plotted (in 1962) from data obtained with neutral-buoyancy floats [65]. Over most of the strait these currents are counterclockwise. However, it has proved difficult to define the position of the zero surface and the water flow below it (at 1200 m of depth). Worthington [70] used for this purpose submerged buoy stations equipped with current and temperature recorders. The cold current had a greater average velocity (46 cm/sec) than the warm one (12 cm/sec). A direct relationship between the velocity of the cold Greenland current and water temperature at many levels was detected.

The temperature and salinity of the Denmark, Faeroe–Iceland and Faeroe–Shetland straits were measured during the second cruise of R/V *Mikhail Lomonosov* [20]. These variables and the vertical stability of water were used to estimate the boundaries of the individual water masses. The main flow of Atlantic water enters the Norwegian Sea through the Faeroe–Iceland Strait. It was shown that the North Atlantic, East Greenland and East Iceland currents interact within the straits.

A comparison of the currents calculated by the dynamic method with direct instrumental observations [21] proved the computed results to not fit reality, probably because the dynamic method is not suitable for the calculation of current velocities in straits with complex flow structure. Observational materials showed that under certain meteorological conditions the main water exchange proceeds via the Faeroe–Iceland Strait.

It might be useful to compare the results of this study with those on the Florida Strait [45], whose dynamics are simpler (there is no horizontal countercurrent). A comparison of the instrumental observations and calculated geostrophic velocities showed an agreement only with simultaneous measurements of temperature and salinity, when the zero surface is specified from special observations. The distance between the horizons of observations must be minimal. Direct observations carried out in the same area [70] confirmed the existence of inertial currents generated by the Earth's rotation. To explain this effect a two-layer flow model, with no momentum transfer from layer to layer, has been suggested.

Until recently, virtually no work was done on the Mozambique Strait. Papers [14] and [37] fill in this gap to some extent. A transverse inhomogeneity shows up in the hydrochemical and hydrological characteristics, and the bend in isotherms and isohalines near the African coast defines the position of the main stream in the Mozambique current.

The application of the dynamic method has proved the existence of a bilateral water exchange in the case of the Vries Strait [6], one of the main straits in the Kurile Islands chain, but the factors responsible for such bilateral currents are not indicated.

In concluding this section, we wish to emphasize that straits with a horizontal water exchange are still not well understood. For example we have no explanation for bilateral currents. The existing observations are fragmentary and random, and recent literature lacks almost any information on such important straits as the Davies and the Bering.

SHALLOW-WATER STRAITS

Many straits connecting small estuaries, lagoons, and bays with the sea (the Kerch and Kinburn in the Black Sea, straits connecting the Gulf of Riga and the Gulf of Bothnia with the Baltic Sea, etc.) belong to this type. The water exchange through shallow-water straits is controlled by river runoff and the wind-actuated surface perturbance (rise or fall) at the ends of the strait. Generally, such straits connect with the sea basins in which fishing or fish raising is important (the Azov Sea, estuaries and bays of the northern Black Sea area), or harbors (Sukhoy firth), river estuaries (the Dnepr estuary, or the Courland Lagoon in the Baltic Sea). The nature of such basins means that the following problems occur in the case of the straits:

1) maintenance of optimal biological conditions in the inland basins in the face of intensified human activity (reduced runoff in connection with preemption of water for power production and irrigation, pollution due to industrial and municipal wastes);

2) protection of potable and industrial water intakes from impression of wedges of saline sea water;

3) the building of barriers protecting the bays and estuaries from tsunami, wind waves, and tidal waves.

Each of these items requires the solution of some specific physical and hydrological problems. The solution of the first one requires knowledge of the dependence of water exchange under natural conditions and after narrowing of the strait by man-made construction. To analyze such relations Lagutin and Tolmazin suggested in 1965 a steady-state model of water flow in a shallow strait with two empirical coefficients allowing for the nonuniform cross section of the strait; Rozengurt et al. [27, 28] extended this method to various straits (the Kerch Strait, the Dnestr estuary). The method gives the change of the average water exchange upon regulation of strait width and length. To control the salinity of the Sea of Azov and the Dnestr estuary, authors suggest partial closure of the strait with a dam cut by long channels for passage of ships and fish. The parameters of the dams capable of maintaining the desired average salinity are given. The calculations have been carefully checked by Altman [1] against his multi-year observations in the Kerch Strait. The background salt level of the Azov Sea was taken into account in his studies. This model could be improved if the strait were considered together with the off- and on-shore areas in the seas connected by the strait. Under these circumstances, the level at the end of the strait would be calculated as a function of the external factors (the wind and river runoff) and water exchange, and would not have to be given from observations. The work of Rozengurt [25] is a step in this direction. Rozengurt analyzed the wind-induced surface perturbations in several estuaries and lagoons to determine the point on the bar closing such lagoons at which water exchange between the estuary and the sea would be most efficient, were a channel through the bar cut at that point.

The data on water exchange are used for the calculation of the components of the local water and salt budget. Rozengurt [24] calculated the expected average salinity of the Dnestr estuary and the time at which equilibrium would be achieved, assuming different runoffs in the regulated river. His calculations were based on an equation for the transient salt budget and a careful application of Lineikin's hypothesis (1955) on the existence of a linear relationship between the average sea salinity and the salinity of the inflowing and outflowing waters (based on the Sea of Azov). A similar method was used by Nitani and Noguchi [56] for calculating the time required for purging a bay from pollutants.

Theories postulating the dependence of salinity of exchanging waters upon the average salinity of the basin were checked against observations carried out over 10 years in Kojima Bay [57], a part of an estuary sealed off from the sea. The water discharge into the sea is strictly controlled by a lock; the sea water penetrates into the bay by filtration. It is shown that the salinity changes calculated from the unsteady-state equation of salinity balance fit the observations. The salt transfer from the lacustrine part into the river estuary part is estimated with the aid of the equation for one-dimensional diffusion.

Regional studies of water exchange through shallow-water straits were carried out to reveal the characteristic features of current dynamics and salt discharge under various conditions [13, 26, 31, 52]. The most detailed investigations are devoted to the water exchange between the Sivash and the Sea of Azov [31], the Courland Lagoon [50] and the Baltic Sea, and are based on long-term observations of currents and other factors. Chapter VII of Bolshakov's monograph [7] presents a calculation with the aid of Tokano's model of the distribution of a "tongue" of fresh water flowing from the mouth of the Dnestr estuary. Tokano's method is based on the determination of the full-flow field in half-space assuming a rectangular river mouth (that is, zero horizontal vorticity of the wind field) at a given water runoff. The density stratification is given in the Reid model. By fitting the coefficients of horizontal turbulent exchange, it is possible to achieve agreement with observations. One of the faults of this method is that the condition of disappearance of the vertical gradient with depth does not hold for the shallow waters of the Dnestr estuary.

The second problem mentioned at the beginning of the Chapter, namely salt wedges, exists in almost all large estuaries and river deltas that have been dredged and where the river runoff has been reduced. Dyer and Ramamoorthly [47] present a detailed description of penetration of saline waters into the Wellar (India) estuary during a tide. As the tide recedes, isolated lakes of saline waters remain at the bottom of the estuary. To model this practical problem, one must examine the process of penetration of saline water wedges into estuaries and their dispersal by turbulent diffusion and wave motion. Hydraulic calculations combined with simulation in model channels proved to be a most efficient method.

The most profound studies of this process were carried out by American engineers and are presented in Chapters VII-IX of the book by Ippen [15, 16, 36] and in Ippen's paper [51]. Thus Keulegan [15] deals with the characteristics of an arrested wedge in a tideless estuary, a watercourse or channel of approach in relation to the velocity of the upper and lower currents and the channel's parameters. The wedge as a whole is immobile. However, there are currents within it. Right below the interface the currents run to the sea, but closer to the bottom the velocity changes sign. The relationship between the wedge length and thickness on the one hand, and the characteristic velocity, the channel parameters, and viscosity coefficients on the other is defined by dimensional analysis. The dimensionless coefficients of the problem were determined by modeling in a channel. The velocity field is examined in a similar manner. Mixing is due to gravitation of instability of internal waves at the interface. The upward salt flow is related by means of empirical correlations to previously computed characteristics of the wedge. Formulas for the calculation of friction forces at the interface and the bottom, based on quadratic dependence of friction on average velocity, are presented. The theoretical calculations and the data for the Mississippi River are in agreement.

Harleman [36] analyzes the frontal collision of saline and brackish waters by means of a one-dimensional, unsteady-state diffusion equation. He reviews the cases of constant and instantaneous salt discharge into the flow. The shift effect is taken into account in the one-dimensional model by means of a diffusion coefficient that is a function of the roughness and the variability of the channel. Most of the paper is devoted to a description of model experiments to determine the exchange coefficient D_t under various conditions. Harleman assumes the Obukhov–Richardson law, that is, $D_t = \mathrm{const}\, \varphi^{1/3} l^{4/3}$, where the rate of dissipation $\varphi^{1/3}$ was determined experimentally in model channels. The dependence of D_t on some other stratification parameters was obtained in a similar way.

Ippen [16, 51] is mainly concerned with the effect of tidal phenomena on the spread of a salt wedge in an estuary. He found that the unsteady-state diffusion of salt can be described by a superposition of a diffusion (quasistationary) part and a transient one, produced by tidal currents. Both phenomena are analyzed by means of a one-dimensional diffusion equation, flow equations for tidal waves of small amplitude, and relations connecting an unsteady shear flow with energy dissipation and diffusion characteristics. The instantaneous salt distribution is related to the tidal conditions in the estuary, the sea water salinity, and the flow of fresh water. An important criterion of the length of the salt zone is derived for estuaries of varying depth.

The problem of preserving the coasts of strait, gulfs, bays, and estuaries from disastrous stormy washes and tides may be solved with the aid of precise hydrologic forecasting consisting of prediction of the possible level of raise of the water following strait damming. This problem is featured in three papers [44, 46, 58]. One of them [46] describes an analysis of tidal waves in terms of a tide model; the other two investigations [44, 58] are based on natural observations and calculations of the level.

In general, shallow-water straits are better understood than straits of other types, primarily because they are more accessible and the research results are of more immediate practical use.

WATER EXCHANGE THROUGH STRAITS AND CLIMATE CHANGES

Theories of climate control are based on the idea that the ocean–atmosphere system is in an unstable equilibrium, which could be disturbed by changing the radiative balance in polar regions. In these theories, straits are not merely morphological narrows but Nature's bottlenecks, in which minor expenditures of energy ought to stimulate spontaneous processes of climatic changes. Many ways for achieving local and global climatic changes are suggested [8, 9, 11, 17]. The most striking project is that of Borisov [9], who advocates climate "melioration." He believes that pumping water from the Arctic Ocean into the Pacific through the Bering Strait would increase the inflow of the warm Atlantic water. This would lead to irreversible melting of the Arctic ice for the simple reason that the "dark" Arctic water would absorb more heat than the light ice does now. However, a qualitative analysis of the chain of anticipated events involved with Borisov's theory of climate "melioration" shows that it fails to take into account the existence of feedback relations within the ocean–atmosphere system which tend to dampen any form of disturbance.

Monin [18] thinks that the system sea–atmosphere is fairly stable to an initial heat shock. Because of that, many projects for controlling the global climate, even though interesting to popular-science readers, are in most cases scientifically unsound. On the other hand, local regulation of conditions in inland seas and of the climate in adjacent coastal areas may be quite feasible.

LITERATURE CITED

The number in parentheses refers to the number of the citation in the Soviet Abstracts Journal (Referativnyy Zhurnal) Geophysics 1970.

1. Altman, E.N., and D.M. Tolmazin, Method for calculating currents and water exchange in the Kerch Strait. Okeanologiya, **10**, No. 3, 1970.
2. Bogdanova, A.K., Hydrologic conditions in the Red Sea and in the Gulf of Aden, according to observations of the R/V *Akademik A. Kovalevskiy*, Okeanolog. issledov. No. 15, Nauka Press, 1966.
3. Bogdanova, A.K., Hydrology of the Bosporus and the pre-Bosporus area of the Black Sea. In: Vodoobmen cherez Bosfor i yego vliyaniye na gidrologiy u i Biologiyu Chernogo morya (Water Exchange through the Bosporus and Its Effect upon the Hydrology and Biology of the Black Sea), Naukova Dumka Press, Kiev, 1969, pp. 5-121 (1970, 12B106).
4. Bogdanova, A.K., V.S. Bolshakov, and D.M. Tolmazin, On the distribution and transformation of the Mediterranean water in the Black Sea. In: Okeanograficheskiye issledovaniye Chernogo morya (Oceanographic Studies of the Black Sea), Naukova Dumka Press, Kiev, 1967, pp. 3-15 (1968, 9B60).
5. Bogdanova, A.K., and D.M. Tolmazin, The mixing of waters of the upper and lower currents in the Bosporus. In: Dinamika vod i voprosy gidrokhimi Chernogo morya (Water Dynamics and Hydrochemistry of the Black Sea), Naukova Dumka Press, Kiev, 1967, pp. 14-25.
6. Bogdanov, K.G., Hydrological conditions in the Vries Strait during the summer months. In: Okeanodogicheskiy issledovaniya (Oceanologic Studies, 1968, No. 19, pp. 95-104 (1969, 6B116).
7. Bolshakov, V.S., Transformatsiya rechnykh vod u Chernom more (Transformation of River Waters in the Black Sea), Naukova Dumka Press, Kiev, 1970 (1970, 9B140).
8. Borisov, P.M., Can man control the climate of the Arctic? Priroda, 1967, No. 12, pp. 63-73.
9. Borisov, P.M., Mozhet li chelovek izmenit' klimat? (Can Man Change the Climate?), Nauka Press, Moscow, 1969.
10. Vinogradov, K.A., and D.M. Tolmazin, Salinity and sea life. Zemlia i Vselennaya, 1968, No. 6, pp. 26-32 (1969, 6B141).
11. Gusev, A.M., Klimati pogoda (The Climate and Weather) Nauka Press, Moscow, 1966 (1966, 6B407).
12. Dobrzhanskaya, M.A., Distribution of oxygen and phosphates in the pre-Bosporus area of the Black Sea. In: Vodoobmen cherez Bosfor i yego vliyaniye na gidrologiy u i Biologiyu Chernogo morya (Water Exchange through the Bosporus and Its

Effect on the Hydrology and Biology of the Black Sea), Naukova Dumka Press, Kiev, 1969, pp. 122-158 (1970, 1B108).

13. Dubra, I. B., Water exchange through the Klaipeda Strait. In: Gidrometerologicheskiy rezhim i agroklimaticheskiye resursy (Hydrometeorological Regime and Agroclimatic Resources) Mintins Press, Vilnius, 1969, pp. 117-123 (1970, 2B115).
14. Yerofeyeva, V. S., and V. B. Rzhonsnitskiy, Some features of the oceanographic conditions in the Mozambique Strait. Trudy Azov o-Chernomorsk n. i. inst. ryb. khoz. i okeanograf. 29, pp. 54-64, 1970.
15. Keulegan, G. H., Mechanism of an arrested saline. In: Wedge Estuary and Coastal Hydrodynamics, A. T. Ippen, ed., McGraw-Hill, 1966, pp. 278-303.
16. Ippen, A. T., Salinity intrusions. In: Wedge Estuary and Coastal Hydrodynamics, A. T. Ippen, ed., McGraw-Hill, pp. 326-356.
17. Mikhaylov, S. V., Ocean and mankind. Economika, 1969.
18. Monin, A. S., Weather Forecasting as a Problem of Physics (translation from Russian), MIT Press, 1972.
19. Osipenko, L. A., Currents in the straits of the Far Eastern seas. Abstracts of the XII scientific conference devoted to 50th anniversary of the Soviet State, Vladivostok, 1967.
20. Paranichev, L. G., and Y. M. Belyakov, Hydrological conditions in the straits of the northeastern Atlantic Ocean during the winter of 1969. In: Morsk. gidrofiz. issledov. No. 4 (46), pp. 229-233, Sevastopol, 1969.
21. Pastukhov, A. F., I. M. Kiryukhina, and A. A. Serebryakova, Water exchange through the Faeroe–Shetland and Faeroe–Iceland Straits. In: Morsk. gidrofiz. issledov. No. 4(46), pp. 220-228, Sevastopol, 1969.
22. Rzheplinskiy, D. G., and V. B. Shtokman, Analysis of circulation around islands in oceans by means of a numeral experiment. Izv. Akad. Nauk SSSR. Fizika atmosf. i okeana 4, No. 12, pp. 1261-1274, 1968.
23. Rozhdestvenskiy, A. V., The inflow and distribution of waters of the Sea of Marmara (Lower Bosporus current) into the Black Sea. Ribno stopanstvo, 16, No. 6, pp. 4-6, 1969 (1970, 7B154).
24. Rozengurt, M. Sh., The currents and water exchange in the Dnestr estuary. In: Okeanograficheskiye issledovaiya Chernogo morya (Oceanographic Studies of the Black Sea), Naukova Dumka Press, Kiev, 1967 (1968, 9B126).
25. Rozengurt, M. Sh., A method of indirect calculation of level perturbation in sea-linked lagoons. Trudy Gos. Okeanogr. Inst., a, p. 98, 1969.
26. Rozengurt, M. Sh., Procedure for observational studies in a strait of a shallow-water basin, based on spectral analysis of current fluctuations. In: Biolog. problemy okeanografii

Yuzhnykh morye (Biological Problems of Oceanography of Southern Seas), pp. 61-63 Naukova Dumka Press, Kiev, 1969 (1969, 8B23).

27. Rozengurt, M. Sh., and D. M. Tolmazin, Theoretical calculation of water exchange between the Dnestr estuary and the Black Sea. In: Dinamjka vod i voprosy gidrokhimii Chernogo morya (Water Dynamics and Problems of Hydrochemistry of the Black Sea), pp. 81-89, Naukova Dumka Press, Kiev, 1967 (1968, 6B128).
28. Rozengurt, M. Sh., D. M. Tolmazin, and B. L. Lagutin, The economic use of shallow-water sea basins. In: Problemy biookeanografii (Problems of Bio-Oceanography), Naukova Dumka Press, Kiev, pp. 14-21, 1967.
29. Seryi, V. V., Seasonal variability of hydrologic conditions in the Gulf of Aden. Trudy Vsesoyuz. n. i. inst. rybn. khoz. i okeanogr. (VNIRO), 64, pp. 117-129, 1968.
30. Seryi, V. V., Water exchange between the Red Sea and the Gulf of Aden. In: Okeanologicheskiye issledovaniya (Oceanologic Studies), Nauka Press, Moscow, 1968, No. 19, pp. 195-200 (1969, 6B112).
31. Slatinskiy, Y. G., Water exchange between the Sivash and the Sea of Azov through the Tonkiy Strait. Sbornik robot Basseynovoy gidrometeorol. observ. Chernogo i Azovskogo morye, 7, pp. 38-54, 1969 (1970, 6B184).
32. Tolmazin, D. M., Steady-state model of water flow in a stratified strait (Bosporus and other straits). In: Okeanograficheskiye issledovaniya Chernogo morya (Oceanographic Studies of the Black Sea), Naukova Dumka Press, Kiev, pp. 18-34, 1967 (1968, 12B98).
33. Tolmazin, D. M., V. A. Shnaydman, and Zh. M., Atsikhovskaya, Problemy dinamiki vod severo-zapadnoy chasti Chernogo morya (Problems of Water Dynamics in the Northwestern Black Sea), Naukova Dumka Press, Kiev, 1969 (1969, 9B98K).
34. Fel'zenbaum, A. I., and N. B. Shapiro, The method of "middle sectioning" and its development. Moorsk. geofiz. issledovaniya, 1969, No. 2 (44), pp. 9-26.
35. Filippov, D. M., Tsirkulatsiya i struktura vod Chernogo morya (Circulation and Structure of Black Sea Waters). Nauka Press, Moscow, 1968.
36. Harleman, D. R. F., Diffusion processes in stratified flow. In: Estuary and Coastline Hydrodynamics, A. T. Ippen, ed., McGraw-Hill, 1966.
37. Khimitza, V. A., Water masses and hydrochemistry of the Mozambique Strait. In: Okeanologicheskiye issledovaniya (Oceanologic Studies), No. 19, pp. 180-188, Nauka Press, Moscow, 1968 (1969, 7B120).

38. Khimitza, V.A., Water exchange through the Bab el Mandeb Strait. Trudy Azovo-Chernomorsk. n. i. inst. ryb. khoz. okeanograf. 1970, No. 29, pp. 3-17 (1970, 8B190).
39. Shtokman, V.B., Qualitative analysis of causes of anomal circulation around oceanic islands. Izv. Akad. Nauk SSSR. Fiz. atmosfery i okeana, **2**, No. **11**, 1966 (1967, B73).
40. Edelman, M.S., Hydrological studies in the southern Red Sea. Trudy VNIRO, **64**, pp. 102-116, 1968, (1968, 11B169).
41. Aagaard, K., and L.K. Coachman, The East Greenland current north of Denmark Strait, Part I. Arctic, **21**, No. 3, pp. 181-200, 1968 (1969, 7B891).
42. Aagaard, K., and L.K. Coachman, The East Greenland current north of Denmark Strait, Part II. *Ibid.*, 1968/1969 (1969, 9B87).
43. Bogdanova, A.K., Particularités de l'échange d'eau par le Bosphore. Rapp. et proc. -verb. réun. Commis. internat. explorat. scient. Mer. Mediter. Monaco, 1969, No. 4, pp. 693–696. (1970, 2B120).
44. Drebs, H., Untersuchung über den Einfluss des Hindenburgdammes auf die Tidenochwasserstände im Wattermeer., Küste, No. 17, pp. 34-50, 1969 (1970, 12B141).
45. Broida, S., Geostrophy and direct measurements in the Strait of Florida. J. Mar. Res., **27**, No. 3, pp. 278–292, 1969 (1970, 10B73).
46. Cowley, J.E., Tidal model of Northumberland Strait. J. Hydraul. Div. Proc. Amer. Soc. Civil Eng., **95**, pp. 827–838, 1969 (1970, 2B136).
47. Dyer, K.R., and K. Ramamoorthly, Salinity and water circulation in the Wellar estuary. Limn. and Oceanogr., **14**, No. 1, pp. 4-15, 1969 (1970, 12B123).
48. Gade, H.G., Horizontal and vertical exchange and diffusion in the water masses of the Oslo Fjord. Helgoländer wiss. Meeresunters., **17**, pp. 462–475, 1968 (1968, 12B149).
49. Heezen, B.G., and G.L. Johnson, Mediterranean undercurrent and microphysiography west of Gibraltar. Bull. Inst. Oceanogr., **67**, No. 1382, 1969 (1970, 4B155).
50. Henriette-Bois, M., Resultats de mesures hydrologiques faites à bord de la *Orygny*, Campagne Mediterranean Outflow, Cahiers oceanogr., **21**, Suppl. No. 1, pp. 49-192, 1969 (1968, 8B107).
51. Ippen, A.T., Salt-water fresh-water relationship in tidal channels. L'énergia elettrica, **6**, pp. 329-333, 1967 (1968, 2B57).
52. Kvinge, Th., On the special current and water level variations to the channel of Säo Sebästäo Biol. Inst. Oceanogr., **16**, pp. 23-37, 1967 (1969, 3B149).

53. Lacombe, H., Resultats des mesures d'hydrologie et de courants effectués à bord de la *Calypso*. Cahiers oceanogr., 21, Suppl. No. 1, pp. 1-48, 1969 (1969, 8B106).
54. Lomniewski, K., Stosunki hydrograficzne w Cieśninach Duńskich. Part I. Zesz. geogr. Wyższa szkola ped. w Gdańsku, 10, pp. 9–35, 1968 (1969, 9B93).
55. Madelain F., Influence de la topographie du fond sur l'écoulement meditéraneen entre le d'etroit de Gibraltar et le cap Saint-Vincent. Cahiers oceanogr., 22, No. 1, pp. 43-61, 1970 (1970, 8B162).
56. Nitani, H., and I. Noguchi, On the pollution in a nearly closed circular harbor. J. Oceanogr. Soc. Japan, 23, No. 3, pp. 126-135, 1967 (1968, 6B174).
57. Okuda, S., On the change in the salinity distribution and bottom topography after the disaster of Kojima Bay. Bull. disas. Prev. Res. Inst., Kyoto Univ., 18, Part 1, No. 132, pp. 35-48, 1968 (1969, 7B166).
58. Preiffer, B., Untersuchung über den Einflüss des geplanten Dammbaues zwischen dem Festlande und der anschliessenden Festlandküste. Küfte, 1969, No. 17, pp. 1-33 (1970, 12B140).
59. Rattry, M., Jr., Some aspects of the dynamics of circulation in fjords. Estuaries, 1967, 45, pp. 52-62, 1967 (1968, B86).
60. Schmitz, W.I., Jr., On the dynamics of the Florida current. J. Mar. Res., 27, No. 1, pp. 121–150, 1969 (1970, 10B72).
61. Seco, S.E., Aplicacion de la ecuacion de Margules al estrecho de Gibraltar Bol. Inst. esp. oceanogr., 1967, p. 132 (1970, 3B134).
62. Siedler, G., Zum mechanismus des Wasseraustauschen zwischen dem Roten Meer und dem Golf von Aden. J. Geophysic., 32, pp. 335-339, 1966 (1967, 9B67).
63. Siedler, G., Schichtungs- und Bewegungsverhältnisse am Südausgang des Roten Meeres. Meteor. Forschungsergeben, A, No. 4, pp. 1–76, 1968 (1970, 2B141).
64. Sternberg, R.W., Friction factors in tidal channels with differing bed roughness. Marine Geology, 6, No. 3, pp. 343-360, 1968 (1969, 2B67).
65. Swallow, J.C., and L.V. Worthington, Deep currents in the Labrador Sea. Deep-Sea Res., 16, No. 1, pp. 77-84, 1969 (1968, 8B112).
66. Svensson, A., Hydrography of the Skagerrak and Kattegat areas. Swedish observations 1967. Ann. biol. Conseil. internat. explorat. mer., 1967, 24, Copenhagen, 1968, pp. 56-57 (1969, 10B88).
67. Thomsen, H., Danish observations in the Great Belt and the Baltic, 1967. Ann. biol. Conseil. internat. explorat. mer., 1967, 24, pp. 57–58 (1969, 10B87).

68. Thomsen, H., Hydrology of the Kattegat area, 1967 Danish observations. *Ibid.*, p. 57 (1969, 10B86).
69. Tolmazin, D. M., Sur la théorie des courants dans les détroits des Mers du sud. Rapp. et proc. -verz. réun. Commiss. internat. explorat. scient. Mer. méditerr. **19**, No. 4, pp. 701-704, Monaco, 1969 (1970, 1B42).
70. Worthington, L. V., An attempt to measure the volume transport of Norwegian Sea overflow water through the Denmark Strait. Deep-Sea Res., **16**, Suppl., pp. 421-432, 1969 (1970, 6B174).
71. Zorefirmanda, M., Water exchange between the Adriatic and the Eastern Mediterranean. Deep-Sea Res., **16**, No. 2, pp. 171-178, 1969 (1969, 10B99).

SEA ICE

V. L. Tsurikov

I. OVERALL PROBLEMS OF SEA-ICE STUDIES

The overall problems of sea-ice research embrace a very great variety of fields. Among such problems are the periodic variations of ice conditions, with a periodicity of two to seven and even more years. The 7-year cycles seem to be closely related with long-period tidal phenomena in the atmosphere [17, 18].

Apart from the 6- to 7-year periodicity of ice conditions there also exists a 4-year cycle, found a long time ago, but still not well understood. We also have a 2-year cyclicity related to changes in atmospheric processes with similar periodicity [13]. Such changes of ice conditions are most distinctly revealed during averaging of multi-year data on ice distribution in even and odd years separately.

It has been noticed that sea-ice fluctuations in the eastern and western parts of the Arctic Ocean adjoining the Soviet Union have opposite phases. This led to a theory of circumpolar "wave" of icing at high latitudes, a wave that propagates from west to east, that is, from Spitsbergen to the meridian of the Wrangel Island.

The problem of interaction of ice and gravity waves also belongs among overall problems. This problem has long been neglected, and recently has some work on the effect of wind and long waves on ice and on the change of wave characteristics upon passage under ice been published. The most comprehensive study of this problem is that of Kheysin [50]. Three papers complementing Kheysin's book were published in 1969 to 1970. Thus Lebedev [28, 29] deals with the unsteady-state propagation of long waves below ice, viewing the latter either as a weightless, ideally flexible film [28], or as a thin elastic plate of constant thickness [29]. Lebedev also deals with the effect of a continuous ice cover such as fast ice and giant ice fields on long waves, and takes into account the viscosity of sea water. He finds that the ice cover greatly increases the wave-damping decrement and reduces wave phase velocity [29].

The third of the papers, that by Bukatov and Cherkesov [10], is somewhat similar to [29]. Again the system is assumed to consist of a finite-depth water layer covered by a thin elastic ice plate. The cylindric rigidity of the plate may be either zero or positive, so that, in essence, one deals with wave propagation under continuous or broken ice. Even though papers [10, 28, 29] do supplement previous work on interaction of ice and gravity waves, the problem is still by no means solved. In particular, we still do not know the changes that will be induced in the spectrum of wind waves upon their passage through a sea covered with ice with given characteristics. Hence, for example it is too early for practical recommendations specifying the particular cases under which ships may use ice as a shelter from storm waves.

II. TECHNIQUES OF SEA ICE OBSERVATION

1. Observations from satellites

Many recent papers deal with the use of the earth satellites for surveying sea ice, specifically with remote television and infrared sensing, as well as interpretation of the resulting ice images. So far, television techniques are more widely used. Among reviews of that subject Bushuyev and Volkov [11] have published a comprehensive summary of data obtained over a period of three years on the results of interpretation of television pictures of ice sent back by meteorological satellites. They find that while theoretically it is possible to obtain a complete daily global coverage of sea ice distribution, in practice the existing cloud cover does not permit this to be done more often than once in 5 days. Soviet ice specialists use specific land areas as landmarks and thus are able to locate, in pictures supplied by Kosmos satellites, specific ice objects with a standard deviation of ±6 km. Such accuracy not only permits mapping of edges and boundaries of ice of various density and age, as well as determination of the positions of individual large floes and icebergs, but also allows one to study ice drift and edge shifts, to define lines of ice divergence and convergence and to estimate the ice concentration. Satellite observations and aircraft reconnaissance supplement each other and thus provide a good record of the ice conditions over large expanses of the ocean. Sinyurin's review [47] covers essentially the same areas as [11].

Uralov et al. [49] used satellite data for analysis of ice-edge shifts in the Barents Sea. Comparison of these data with air reconnaissance data and ship observations has shown there are practically no discrepancies in estimates of ice-edge positions. Knapp [73] concluded from information obtained by the Nimbus satellite that open water areas in the Antarctic are much more numerous and larger than previously thought. Sissala [96] used the same data to trace the movement of one of the giant tabular icebergs in Antarctic waters.

Barnes et al. [63] describe ice observations from the Nimbus-2 satellite, equipped with high-resolution infrared cameras, which allow observations regardless of the available light. These authors discuss the possibility of mapping ice boundaries from such observations, and compare the results of aerial photography and infrared studies.

Finally, it is worth mentioning papers [71, 97], which among other oceanographic observations, discuss ice studies from satellites. These studies should be important to those ice specialists who are interested in studies of ice as a function of other factors, such as sea state, currents, or temperature.

2. Aerial reconnaissance

Over the last decade, airborne instrumental ice surveys have progressively superseded visual reconnaissance. Among the new techniques radar observations, which are also useful during the Arctic winter darkness, have become important. It is characteristic that in the period 1969-1970 there were three papers dealing with radar reconnaissance and not a single one on visual methods. Thus, radar observations from aboard a helicopter [83] are described. In this case the object was to observe separation of icebergs from Store glacier (Greenland). The two other papers [92, 93] describe methods for estimating the age of polar ice packs from their appearance on the screen of the plane's radar which operates at the wavelength of 2.25 cm. It is pointed out that in the presence of low cloud cover, no aerial photography, whether in the visible or the infrared, can be performed, whereas radar yields excellent results. The author derived an expression for the scatter coefficient of the radar energy as a function of the angle of incidence. The equation incorporates a dimensionless coefficient which depends on the age of ice that acts as the radar reflector. This coefficient varies from 0.08 in the case of ice-free water and new ice to 0.35 to 0.65 in the case of polar ice.

3. Observations from the shore

Radar is now widely used by shore-based ice observation stations. It appears that radar is used in this way only in the USSR, since all four papers that have appeared on this subject were written by Soviet oceanographers. Thus Moyseyev [32, 33] describes shore-based radar observations of ice drift and data interpretation. He indicates factors which tend to lower the possibility of finding passive reflectors, describes methods for interpretating data obtained for seas with regular semidiurnal tides, and for the discovery of tidal and nonperiodic drift.

Moyseyev [33] also describes the interpretation of information on ice drift obtained with the aid of a perspectometer designed by Ivanov. Moyseyev was able to compute the nonperiodic drift component, the wind coefficients for various sites on the White Sea shore, and ice discharge through straits.

4. Ice-based observations

Sampling of ice for the purpose of analysis of its physical and chemical properties is a laborious and complicated procedure. Until recently it required several hours of work and was responsible for distortion of measured ice properties because the ice meanwhile lost part of its brine. For this reason, the annular ice corer described by Cherepanov [58] deserves serious attention. This device produces relatively quickly vertical cores 3 to 4 m long and as much as 320 mm in diameter.

Twiss [98] briefly described recent subice studies in the Antarctic, employing the Scuba apparatus. The latter consists of a camera lowered below the ice, which is linked to the surface by a three-section telescope tube. It was used to measure ice growth.

5. Ice nomenclature

The development of ice terminology or nomenclautre is an integral part of ice research because it is important that the same terms be used everywhere to denote the same ice phenomenon.

Volkov and Treshnikov [14] recently published a Russian translation of the "International Sea Ice Nomenclature" adopted by the World Meteorological Organization. The latter nomenclature is close to the ice terminology used in the USSR until 1956 [21] and is much more complete than the previous approved international nomenclature [2].

Grosvald et al. [16] produced a new classification of shelf ices, greatly expanding the terminology of ice types that are transitional between sea and the glacier types.

III. STUDIES OF ICE DRIFT AND CONCENTRATION

1. Tidal and wind-induced ice drift

The relative importance of tidal ice drift is still unknown in the case of several areas of the ocean that freeze in winter.

Underestimation of this effect leads to serious errors in drift observations and data interpretation. This fact was apparently taken into account by Ingram et al. [72] who observed the drift of a special radio-beacon-equipped buoy in the ice of the Gulf of St. Lawrence. The beacon position was determined from airborne observations, and the data working indicated that drift with diurnal and semidiurnal periodicity is an important factor.

It is now generally acknowledged that wind-induced ice drfit cannot be analyzed independently of currents caused by this drift, owing to friction at the ice–water interface [43]. Bearing this in mind, Yegorov [22, 25] developed a theoretical model of ice drift and under-ice currents, assuming horizontally heterogeneous wind field. In one of his papers [22] he analyzes gradient currents produced under drift ice by sea-level slope. Such currents appear not to affect the wind-induced drift velocity, which contradicts the data of [43], but deflect the ice to the right of the path over which it would have drifted in the absence of gradient currents. The magnitude of this deflection effect is inversely proportional to the sea depth. Yegorov finds that the ratio of the calculated drift velocity and the surface wind is 0.05; that is it is much larger than the observed ratio (which is 0.02 for the Arctic, and 0.03 for the northern part of the Caspian and Azov Seas).

In the other paper [23], Yegorov analyzes the drift of ice floes whose horizontal dimensions exceed their thickness by many times. He starts with the unsteady-state equations for ice movement, and takes into account the Earth's rotation, friction on ice both surfaces, and interaction of ice floes as described by Laykhtman. Assuming definite values for the respective variables, Yegorov shows that ice drift becomes stabilized in 3 hours afters its beginning. From that point on, Yegorov's theory assumes that the steady-steady equation of drift motion and the nonlinear theory of turbulence in boundary layers above and below the ice cover apply. The thus predicted drift of the Severnyy Polyus-4 station proved to coincide with the station's actual course.

Ingram et al. [72] have also compared the actual ice drift with calculations based on the Reed and Campbell theory, which allows for friction at the ice–air and ice–water boundaries, as well as for the Coriolis force. Their calculated data were compared with the drift of the above-mentioned radio beacon in the Gulf of St. Lawrence. Ingram et al. conclude that the theoretical model needs to be improved.

The analysis of the subice drift current, based on observations in the Arctic Ocean, was published by Belyakov [6]. Nikolaeva [34] also analyzes empirical studies, in her case the 10-year ice drift observations in the East Siberian Sea. She treats the drift as a sum of motion components that are parallel and perpendicular to the coast. The ratio of these components depends on the distance to the fast-ice edge and increases away from that edge. In winter, this effect is noticeable at a distance of 1000 km, while in spring it is significant to 400 km.

2. Ice floe rotation in drift and ice concentration

Timokhov [46, 47] analyzed theoretically the ice-floe rotation in drift, which is of importance in analyses of the changes of radar images of floes, in some acoustic problems, etc. Timokhov concludes that ice-floe rotation is controlled by changes in direction of the wind, currents, and ice-drift velocity. The friction of drift ice floes against fast ice is not examined. The angular velocity of ice-floe rotation increases with the decrease of drift velocity and ice-floe radius. The ratio of the turbulent and mean movement of the ice is determined by the water resistance to its movement, the current velocity directly beneath ice, and the Coriolis force. The relationships thus derived hold for the movement of an individual ice floe not touching other floes, but only approximately for real ice covers. Still, special 4-hour observations gave results close to the theoretical expectations.

Doronin [20] has published a theoretical solution to the problem of ice concentration under wind-drift conditions. The interaction between ice floes is taken into account in terms of changes of their momentum. Doronin assumes that the effective factor of friction of ice floes against each other is directly proportional to concentration. He also assumes that the coefficient of vertical turbulent exchange of momentum varies linearly in a certain layer, whereas above this layer the wind velocity is geostrophic. The change of concentration in time is taken as equal to the divergence of the product of velocity and concentration. Doronin also allows for the change of concentration because of melting, using Zubov's relations. The numeral solution is obtained on a time grid with a 24-hour interval. Some of the parameters were taken from observations. The calculations show that concentration changes more rapidly in cases when ice floes fail to contact each other. This is well-known from observations.

IV. STRUCTURE, PHYSICAL AND CHEMICAL PROPERTIES OF SEA ICE

1. Structure and chemical properties of ice

The structural studies of sea ice published in the last 12 years have dealt mainly with pure ice samples consisting of vertical plates (crystallates) that are parallel to each other and constitute large formations known as crystallites (terms used by Davrov [27]). Since the crystallates are parallel, crystallites display similar optic and some other properties. Vertically elongated cylindrical pockets filled with brine constitute the interstitial space between crystallites.

This structure of sea ice determines many of its properties, such as density, thermal properties, modules of elasticity, and stress resistance. The most comprehensive review of structural research on sea ice was published by Weeks and Assur [104]. Later, Weeks and Lofgren [80] described some details of formation of such sea-ice structure as a function of growth rate, obtained from laboratory experiments involving freezing of a sodium chloride solution. Tsurikov [56] had solved the same problem earlier, on the basis of his own laboratory experiments and assuming the model of sea-ice structure. Shesteperov [57] used observational data collected by Johnson on chlorinity of ice and waters of Gullamar Fjord (western Sweden) to confirm Tsurikov's formula.

Lake and Lewis [75] improved the above model of sea-ice structure and showed that the cylindrical brine inclusions have lateral branches. Hence, it follows that aside from the well-known vertical migration of the brine, there exists horizontal migration. Lake and Lewis also proved that there exists convective water flow in the lower layer of growing ice, specifically in the still not fully closed pockets with vertical dimensions of several centimeters. They conclude from this that the water temperature immediately below the ice should exhibit a periodic variability, and that the growth of ice itself is not continuous but proceeds in discrete steps.

And, finally, Hosai [53, 54] studied the chlorophyll content in Antarctic sea ice and in the under-ice water of the Shiowa base.

2. Mechanical properties of sea ice

The determination of mechanical properties of natural ice is rather difficult. Measurement of several of such properties directly within the ice cover requires equipment that is difficult to use under field conditions. Hence, the only properties that have been measured on a substantial scale are the modulus of elasticity and bending strength. For this reason, any new technique capable of direct measurements or even indirect evaluation of mechanical properties of ice cover is of major practical interest, especially if it is a technique consisting of observations from an airplane or a helicopter. Among such new techniques it is worth mentioning that described in [90, 91], whereby the dynamic ice hardness is determined from the depth of penetration of a falling weight. On the basis of experiments with artificial and natural sea ice, Ross [90] defined the optimal shape of the weight and derived a model of weight penetration into ice. He concluded that the critical rate of weight penetration can be calculated provided one knows the temperature and salinity of ice and assuming that the above described structure of the ice is valid [104]. The computed rates agree well with observations. Kheysin and Cherepanov [52] investigated the ice-structure changes at points of weight impact and found that these changes do not depend on the impact velocity or the temperature of the ice cover.

Smirnov [42] approaches the problem of ice strength from another angle. He maintains that ice disintegration as such is not an adequate measure of its susceptibility to breakage by ice breakers, and instead suggests estimating the potential resistance of the ice, that is, the product of ice thickness and its mean vertical bending strength. This property is calculated from the visually observable age and rate of ice decay or from the average 10-day air temperature and height of the ice-covering snow.

The methods for measuring various mechanical properties of ice on small samples are much more advanced. However, the problem here is that extraction from a natural ice cover distorts the sample properties owing to abrupt temperature changes, brine loss, and also because of the scale effect (that is, the smaller the ice sample, the larger its measured bending strength, reduced to this is standard sample size). Lavrov [27] explains the scale effect by assuming constancy of absolute elongation of those ice layers which are under tension during bending. Hence, the sagitta becomes directly proportional to the square root of the beam length. And since the sagitta is linearly related to the bending strength σ, we have

$$\sigma_2 = \sigma_1 l_1^{1.5} h_2 / l_2^{1.5} h_1,$$

where h_1 and h_2 are the heights of two beams of square cross section, while l_1 and l_2 are the beams lengths. By using this expression, Lavrov is able to reduce σ to standard ice-sample sizes.

Lavrov also recommends to use the so-called "structurally modeled (or similar) ice" for the study of ice properties. The crystal structure of a sample of this ice resembles the structure of the ice cover, but on a smaller scale. Lavrov examines, on such models, various cases of ice behavior under load, with due allowance for the rate of load application.

Kheysin [51] analyzes the elastic properties of polar sea ice in connection with his analysis of propagation of flexity and gravity waves under the ice cover. The latter is represented as an isotropic elastic plate floating in an ideal, homogeneous incompressible fluid of known depth. Schmidt [94] explores the problem of disintegration of the ice cover in the Arctic ocean under wind-induced bending using his own observations of ice cracks in the central Arctic ocean.

3. Acoustic properties of sea ice

The acoustic properties of sea ice are being studied from two points of view. First, measurements of propagation of elastic vibrations, as well as their intensity and frequency, are used to determine the site, force, and duration of hummocking. Secondly, the acoustic properties of ice proper are studied in connection with the problem of hummocking and for purposes of estimating the accuracy of large-scale ice-thickness measurements done from submarines or from moving sleds with the aid of sonar devices.

The progress in the hummocking problem is summarized in a review [74] of the results of acoustic research carried out by U.S. drift stations in the Arctic area. Developments in the second area are described in [76, 89]. These papers deal with studies of natural and artificial sea ice. If the receiver is 1.3 to 7.6 m away from the emitter, the attenuation of vibrations at frequencies <2.5 kHz is very low. At higher frequencies the attenuation is greater and at 150 kHz approaches 6 dB/m. It then sharply increases and at 200 kHz reaches 54 dB, probably because at these frequencies the wavelength approaches the transverse size of the ice crystallites.

4. Thermal properties of sea ice

Ono [35] has published his fourth paper in a series of communications on thermal properties of sea ice. In this paper his previous model of sea-ice structure is supplemented by assumptions on the presence of gaseous inclusions in the form of vertical (parallel to heat flow) chains of spherical bubbles. Ono then calculates the density, thermal conductivity, heat of melting, and heat capacity of the ice. He then calculates the diffusion of heat through the ice. The calculations were checked experimentally with thermocouples frozen into ice of the La Perouse Strait, whereby containers filled with cooling mixture were ice alternately placed over or removed from the thermocouples. The recorded temperature changes coincided with the theoretically predicted changes.

Brown [65] collected interesting data on the thermal conductivity of sea ice by measuring heat distribution in the ice cover of the Bering Strait. There are no objections to the upper limit of thermal conductivity ($6 \cdot 10^{-3}$ cal/cm · deg · sec) that he found. The lower limit is a long-disputed matter (see [57]). Some workers think that the lower limit is close to $2 \cdot 10^{-3}$ and even to $1 \cdot 10^{-3}$ cal/cm · deg · sec, whereas others insist it cannot be lower than $4 \cdot 10^{-3}$ cal/cm · deg · sec. Brown obtained values ranging from $3.89 \cdot 10^{-3}$ to $6 \cdot 34 \cdot 10^{-3}$, in other words, seemingly supporting the second viewpoint. However, Brown himself remarks that the rate of ice growth calculated with these coefficients is too high, and that better results are obtained with the smaller conductivities of Schwerdtfeger.

5. Electrical properties of sea ice

The study of electrical properties of sea ice, which was started in the USSR in 1934-1935, had not drawn much attention. Investigations were renewed only in 1959-1960, probably in connection with attempts to develop a method for measuring ice thickness from the rate propagation of electromagnetic waves, using an airborne emitter and receiver [57].

Present studies deal with the dielectric constant ε, resistance ρ (or its reciprocal, that is electroconductivity $1/\rho$) the loss angle tangent tan δ, and the relaxation time, i.e. the time in which ice polarization is attenuated from its initial value to e^{-1}. Thus Addison [61, 62] studied the complex dielectric constant $\varepsilon = \varepsilon' - i\varepsilon''$ on an artificial sea-ice sample and the relaxation time τ. Here ε' is the effective permeability and ε'' is the dielectric constant with allowance for loss, while $i = \sqrt{-1}$. The ice temperature and salinity in Addison's experiments ranged from -12.5 to -35° and 4 to 20‰, respectively, and the frequencies ranged from 20 Hz to 100 MHz. Three ranges of dispersion of ε and τ were found: a low-frequency range (20 to 5000 kHz), an intermediate frequency range (5 to 500 kHz), and a high-frequency range (0.5 to 100 MHz). The first is due to generation of a volumetric charge in the cylindrical brine pockets. Variables ε' and ε'' greatly depend upon temperature and with its change from -15° to -35° they drop by two orders of magnitude. At frequencies below 1 MHz, tan $\delta \approx 10$, while above 10 MHz, it is less than 1. The curves of electrical properties show abrupt inflection, at temperatures of -23° to -24°, i.e., close to the eutectic point of NaCl solutions.

Bogorodsky and Khokhlov [7] analyzed the ε' and ρ of natural 1-year and multi-year sea ice. They have also found the above-cited curve inflections at the eutectic point of NaCl solutions. Their values of ε at low frequencies were of the same order as those observed by Addison.

6. Radiative properties of sea ice

Timerev [44, 45] has investigated the radiative budget of snow and ice in the central Arctic and found that on the average 88% of the incident radiation penetrates into leads and only 68% into melting puddles, 50% into new ice, 58% into the dry polar ice. But 3 to 5 times less radiation passes through old, dry polar ice than through new ice [44]. Langleben [77] found that with increase of puddling the albedo of melting sea ice decreased linearly from 49% to 19%.

V. GROWTH OF SEA ICE, ITS THICKNESS AND MELTING

1. Theoretical study of ice growth

The theory of growth of ice thickness was suggested by J. Stephan as far back as 1891. All later studies are merely supplementary or refinements of Stephan's theory. Most of the later theory is equally applicable to fresh-water and sea ice, with only parameters of theoretical equations depending upon salinity. Three papers on this subject appeared in the period 1969-1970.

Lock et al. [79] investigated the continuous and periodic growth of ice at the lower surface. In the first case, the theoretical analysis is fairly simple, whereas in the second a simplification, namely linear change of temperature with depth, had to be assumed to permit a solution. Lock gives an analytical (based on the perturbation method) and numerical solutions (assuming discrete levels of the three-dimensional variables in the latter case). The results with either approach differ by 16% from the experimental data. In his other paper [78], Lock explores the applicability of asymptotic solutions to problems involving the freezing and melting of ice at its lower surface, with allowance for temperature, heat flow, convection, and radiation. The neglect of boundary conditions had previously sharply restricted the application of solutions. Lock estimated the errors and the order of magnitude of the terms in the newly derived nonlinear differential equations which can be solved either graphically or numerically. The advantage of this calculation method is applicability to all ice-water boundary problems.

Foster [68] checked the theory of generation of convection cells on the lower surface of growing sea ice in the laboratory. As we stated above, such cells are assumed to be the basic structural elements of the ice cover [104]. Cells may be generated due to cooling or salinization of water at the ice surface. Foster found that at water salinity of less than 24.7 ‰ there is no heat convection, whereas above 24.7 ‰ the heat convection has a pronounced effect on the salt convection. Hence, the development of heat convection depends on salinity. Horizontal cell dimensions in salt convection proved to be close to those calculated theoretically (2 to 3 mm).

2. Calculation of ice growth in various sea basins

Baranov and Nazintsev [4] derived an expression for the equilibrium ice thickness in the Arctic Ocean, whereby summer melting equals the winter growth. They took into account ice drift and the heat in flow from the water. A very similar problem was solved by Shesterikov [60] for the same sea. He calculated by the method of consecutive approximations the time interval over which the ice that was heated during the Arctic summer again cools to the point at which freezing of under-ice water resumes and ice growth recommences. The calculated time intervals proved to be very close to those observed.

Untersteiner and Maykut [100] suggest a more complicated method for calculating the thickness of Arctic drift ice. They allow for the radiation budget, turbulent heat exchange with air and water, and atmospheric precipitation. The increase of ice thickness through rafting and hummocking is neglected. The height of the ice snow cover and the distribution of temperature within the cover are computed by numerical integration on a computer. Average multiyear data are input parameters. Untersteiner and Maykut obtained the same equilibrium ice thickness of 3 m as Baranov and Nazintsev [4]. They have also calculated the possible changes of ice conditions with climatic changes, as well as changes of oceanographic conditions, such as lack of snow, cessation of inflow of Atlantic water, lower albedo, etc. However, Untersteiner and Maykut make it clear that it is still impossible to foresee all the possible consequences of artificial melting of ice in the Polar Basin.

A study devoted to the prediction of ice growth to 20 to 25 cm at the eastern Arctic coast [19] based on observed dates of stable ice formation, must also be mentioned.

3. Statistical calculations of ice thickness distribution. Ice melting

Statistical studies of ice thickness based on synchronous observations have been published in recent years. These studies are of interest because even in those areas where the ice is not deformed, its thickness is very variable due primarily to the nonuniformity thick overlying snow cover. Romanov [36] finds that the distribution of the fast-ice thickness in the area of Mirnyy is almost Gaussian. Buzuyev and Shesterikov [9] have statistically analyzed the distribution of fast-sea-ice thickness with allowance for its hummocking, and suggested the use of a factor equal to the ratio of the thickness of ice of given degree of hummocking to the average thickness of smooth ice. The data thus calculated permit one to predict the average ice thickness at the given instant of observation, at the end of ice growth, the time of fast ice formation, probable ice melting, etc.

The decrease of ice thickness by melting is computed by similar methods. This is not surprising, for so far there is no general theory of ice melting such as exists for ice growth. Romanov [37] has statistically analyzed the thickness of drift ice in the eastern Arctic for different magnitudes of ice-cover decay. Kozlovskiy [24] describes the results of experimental studies of ice melting in sea water, whereby ice samples were cut from icebergs and submerged to different depths. Kozlovskiy found that in near-bottom water ice melts more rapidly than at the surface, which leads him to conclude that the subsurface parts of icebergs and ice shelfs do melt all year round.

The heat balance in ice is intimately connected with its growth and melting. Attempts to evaluate the areal heat-balance distribution in sea ice have started long ago and have been renewed lately. Thus Brown [65] calculated the heat budget of ice in the Bering Strait, based on 4-day observations separated by 12 to 29 days. Untersteiner [99] published a program of heat-budget research in the Arctic.

VI. APPLIED STUDIES OF SEA-ICE COVER

The topics covered by sea-ice research are fairly varied. Most of the studies are devoted to problems of shipping in polar seas, particularly the possibility of navigation under various ice conditions [8, 31, 38, 39, 66], the economics of shipping under various ice conditions [95], and the experience gained in Arctic navigation [26]. Among other practical matters discussed special attention is paid to the problem of the icing of ships [40], the exploitation of the Arctic ship route [69], to a special device for measurement of pressure on oil rigs in Cook Inlet (Alaska) [84] and to the problem of polluted ice produced by a crash of an American plane at Thule, Greenland [70].

VII. REGIONAL STUDIES OF SEA ICE

The subjects of regional studies vary greatly. There are many publications of ice data, mainly related to the north Atlantic and adjoining seas [82, 83, 101, 102] and to waters of the American

Arctic [64]. A comprehensive description of ice behavior in the sea of Okhotsk [1], a symposium on sea ice in Icelandic waters [86], as well as several papers on ice distribution and ice cover in the North Atlantic and adjacent seas [30, 48, 55, 67], the Baltic Sea [15] and the Antarctic [87] also belong in that group, as does a paper on the effect of ice on bottom sediments in the Sea of Okhotsk [3]. A more detailed analysis of applied and regional investigations of sea ice is beyond the scope of this review.

LITERATURE CITED

The number in parentheses refers to the number of the citation in the Soviet Abstracts Journal (Referativnyy Zhurnal) Geophysics 1970.

1. Akagawaa, M., Ice of the Sea of Okhotsk. Kisho Kenkyu Noto, 1969, No. 101, pp. 449-479 (10B133).
2. Al'bom aerofotosnimkov ledovykh obrozovaniy v more (Album of Aerial Photographs of Sea Ice Formations). Gidrometeoizdat Press, Leningrad, 1960, 223 pp., 140 photographs.
3. Archikov, Ye.I., The effect of sea and river ice covers on the deposition of bottom sediments near Shantar Islands. Zapiski Zabaikal. fil. geogr. obshch. SSSR. No. 40. pp. 87-89 (8B196).
4. Barinov, G.I., and Yu.I. Nazintsev, A simplified model of equilibrium state in drift ice of the Polar Basin. Numerical calculations. Trudy Arkt. i Antarkt. n. i. inst., 1970, No. 291, pp. 32-43 (10B131).
5. Belyakov, L.N., Summer freshening of surface sea waters, the dynamics of drift currents and ice movement in the Polar Basin. Okeanologia, IX, No. 3. 1969 (2B42).
6. Belyakov, L.N., Some peculiarities of under-ice drift currents in ice-covered sea. In: Probl. Arktiki i Antarktiki, No. 33, Gidrometeoizdat Press, Leningrad, 1969, pp. 30-38, (11B61).
7. Bogorodskiy, V.V., and G.P. Khokhlov, Electrical characteristics of Arctic drift ice at frequencies of 11 Hz to 1 MHz. Dokl. Akad. Nauk SSSR, **189**, No. 6, pp. 1230-1232, 1969 (7B138).
8. Buzuyev, A.Ye., and A. Ya. Ryvlin, Calculated resistance of ice to icebreaker progress under various ice conditions. In Probl. Arktiki i Antarktiki, No. 31, Gidrometeoizdat Press, Leningrad, 1969, pp. 69-73.

9. Buzuyev, A. Ya., and N. P. Shesterikov, The dependence of average fast ice thickness on hummocking. *Ibid.*, No. 32, 1969, pp. 30-36 (2B106).
10. Bukatov, A. Ye., and L. V. Cherkesov, Unsteady fluctuations of ice cover cuased by periodic shifts of pressure. In Morsk. gidrofizich. issledovaniya, No. 2(44), Sevastopol, 1969, pp. 94-105 (5B73).
11. Bushuyev, A. V., and N. A. Volkov, Meteorological satellites for ice observation. In: Probl. Arktiki i. Antarktiki, No. 33, Gidrometeoizdat Press, Leningrad, 1970, pp. 5-12 (10B30).
12. Vaygachev, A. Z., and N. F. Tretiakov, Shift of the Pobeda ice island. Inform. Byull. Sov. Antarkt. Eksped., 1970, No. 76, pp. 57-59 (9B157).
13. Volkov, N. A., and B. A. Sleptsov-Shevlevitch, The two-years cycle of ice-cover fluctuations. In Probl. Arktiki i Antarktiki, No. 34, Gidrometeoizdat Press, Leningrad, 1970, pp. 13-19 (10B132).
14. Volkov, N. A., and A. F. Treshnikov, New international sea-ice nomenclature. *Ibid.*, No. 32, pp. 132-143, 1969 (2B108).
15. Vorsulayev, V. G., and K. P. Shirokov, Characteristics of hydrological year and Baltic Sea ice cover during the winter of 1968/69. In: Collection of Papers of the Leningrad and Petrozanodsk Hydrometeorological Observatories, 1970, No. 6, pp. 85–154 (10B140).
16. Grosvald, M. G., V. M. Kotlyakov, and P. A. Shumsky, Shelf ice: definition, terminology, classification. In: Antarktika, (Antarctic), Nauka Press, Moscow, 1969, pp. 114-121 (6B485).
17. Gudkovich, Z. M., E. I. Sarukhanyan, and N. P. Smirnov, Polar tide in the atmosphere at high latitudes and ice cover fluctuations in Arctic seas. Dokl. Akad. Nauk SSSR, **190**, No. 4, pp. 954-957, 1970 (7B144).
18. Gudkovich, Z. M., E. I. Sarukhanyan, and N. P. Smirnov, Polar pressure tide and its effect on the ice cover over the Arctic Seas, Okeanologiya, **X**, No. 3, pp. 426-437, 1970.
19. Gushchenkov, Ye. M., A method for forecasting the dates of achievement of ice thickness of 20 to 25 cm. In: Probl. Arktiki i Antarktiki, No. 32, pp. 45-51, 1969 (2B107).
20. Doronin, Yu. P., A method for calculating ice concentration and ice drift. Trudy Arkt. i Antarkt. Inst., **291**, pp. 5-17, 1970 (10B67).
21. Yevgenov, N. I., Al'bom ledovykh obrazovaniy na moryakh (Album of ice formations at sea). Gidrometeoizdat Press, Leningrad, 1955, 18 pp., 107 photographs, 4 charts and schemes.
22. Yegorov, K. L., Some problems of theory of wind-induced ice drift. Trudy Leningr. gidrometeorol. inst., No. 37, pp. 41-50, 1969 (7B77).

23. Yegorov, K. L., The theory of ice field drift in a horizontally inhomogeneous wind field. In: Probl. Arktiki i Antarktiki, No. 34, pp. 71-78, 1970 (10B69).
24. Kozlovskiy, A. M., Melting of continental ice in sea water. Inform. Byull. Sov. Antarkt. Eksped., 1969, No. 75, pp. 38-42, (5B103).
25. Kozlovskiy, A. M., Fast ice formation in the Gulf of Alasheyev. *Ibid.*, 1970, No. 76, pp. 49-51, (9B156).
26. Kupetskiy, V. N., Scientific and operational support of navigation in the eastern Arctic in 1967. Izv. geogr. obshch SSSR, **102**, No. 1, pp. 66-68, 1970 (8B227).
27. Lavrov, V. V., Deformatsiye prochnost' l'da (Ice Deformation and Strength). Gidrometeoizdat Press, Leningrad, 1969, 206 pp., (3B462K).
28. Lebedev, A. N., The effect of ice cover on the distribution of surface gravity waves in a viscous fluid. In: Morskiye gidrofizich. issledovaniya, No. 2 (44), Sevastopol, 1969, pp. 121-136, (5B72).
29. Lebedev, A. N., The effect of viscosity on the distribution of surface waves in an ice-covered sea. In: Probl. Mirovogo okeana (Ocean Problems), Moscow Univ. Press, 1970, pp. 167-170 (10B84).
30. Lebenev, A. A., The probability of encountering ice in the Barents Sea, the Sea of Greenland, and in the North Atlantic. Trudy Polar. n. i. i proyektn. inst. Morsk. i rybno. khoz. i okeanografii, No. 23, pp. 138–142, 1968 (1B157).
31. Lebenev, A. A., Icebreaker searches for the Northwest Passage. Nat. Fisherman, **50**, No. 3, B-20, 1969 (2B162).
32. Moiseyev, V. S., Results of radar ice-drift observations in the White Sea. Papers of Arkhangel. hydrometeorol. observ., No. 7, pp. 89-93, 1970 (8B143).
33. Moiseyev, V. S., The possibility of forecasting ice drift. Papers of Arkhangel. hydrometeorol. observ., 1970, No. 7, pp. 82-88 (9B113).
34. Nikolayeva, A. Ya., The effect of fast ice upon ice drift. In: Probl. Arktiki i Antarktiki, No. 33, Gidrometeoizdat Press, Leningrad, pp. 110–114 (11B63).
35. Ono, N., Thermal properties of sea ice. IV. Thermal constants of sea ice. Teion Kagaku. Physics Section. 1968, No. 26, pp. 329-349.
36. Romanov, A. A., The distribution of fast-ice thickness in the area of the Mirny observatory. Inform. Byull. Sov. Antarkt. Eksped., 1969, No. 75, pp. 43-47 (5B143).
37. Romanov, A. A., The average thickness of drift ice in eastern Antarctic waters. *Ibid.*, 1969, No. 73, pp. 22-26 (4B142).

38. Romanov, A. A., The effectiveness of a channel opened up in the fast ice near Mirny and optimal time for its passage. *Ibid.*, 1970, No. 76, pp. 30-36 (9B179).
39. Romanov, A. A., The ability of the icebreaker *Ob* to pass through sea ice. *Ibid.*, 1969, No. 75, pp. 48-52 (5B177).
40. Sawada, A., The icing of ships in the Sea of Okhotsk. Kisho Konkyu Noto, 1969, No. 101, pp. 492-497 (9B180).
41. Sinyurin, Yu. N., Using satellite-provided ice information. Meteorol. i gidrolog., 1969, No. 8, pp. 94-96 (1B33).
42. Smirnov, V. I., The potential resistance of ice, from aerial and ship observations of the sea ice cover. In: Probl. Arktiki i Antarktiki, 1969, No. 33, pp. 115-118, Gidrometeoizdat Press, Leningrad (11B188).
43. Smirnov, V. I., Sovetskaye Arktika (Soviet Arctic Seas and Islands of the Arctic Ocean). Nauka Press, Moscow, 1970, 526 pp.
44. Timerev, A. A., Penetration and absorption of radiation by drift ice in the Arctic. Trudy Arktiki i Antarktik n. i. instit., 287, pp. 188-202, 1969 (6B188).
45. Timerev, A. A., Natural illumination and penetration of light into the snow of the central Arctic. In: Probl. Arkt. i Antarktiki, No. 34, Gidrometeoizdat Press, 1970, pp. 35-41 (9B408).
46. Timokhov, L. A., The relationship between the turbulent and mean ice movement. *Ibid.*, Antarktiki, Leningrad, No. 34, 1970, pp. 63-70 (10B68).
47. Timokhov, L. A., The theory of ice-floe rotation in ice cover of varying concentration. *Ibid.*, No. 33, 1969, pp. 46-56, (11B62).
48. Uralov, N. S., Nature and causes of East Greenland coastal sea ice cover variations. Trudy Gos. okeanografich. inst., 1969, No. 96, pp. 3-38 (12B89).
49. Uralov, N. S., A. A. Lebedev, V. S. Loshchilov, and V. A. Voyevodin, Patterns of shift of the ice edge in the Barents Sea. Papers of Murmansk Hydrometeorol. Observ., 1969, No. 2, pp. 36-44 (8B140).
50. Kheysin, D. Ye., Dinamike ledyanovogo pokrova (The Dynamics of the Ice Cover). Gidrometeoizdat Press, Leningrad, 1967, 215 pp.
51. Kheysin, D. Ye., The distribution of bending and gravity waves in an ice-covered sea. Izv. Akad. Nauk SSSR. Fiz. atm. i okeana, 5, No. 12, p. 1334, 1969 (3B86DEP).
52. Keysin, D. Ye., and N. V. Cherepanov, Change of ice structure in the zone of impact of a solid on the ice surface. In: Probl. Arktiki i Antarktiki, No. 34, Gidrometeoizdat Press, Leningrad, 1970, pp. 79-84 (9B403).

53. Hosai, T., Ecological study of colored sea ice layer at Shiowa Nan Kyoku Shiryo, 1969, No. 34, pp. 60-72 (12B116).
54. Hosai, T., Seasonal changes of α-chlorophyll and hydrological conditions under ice in the Shiowa base area. *Ibid.*, 1969, No. 35, pp. 52-67 (5B141).
55. Khrol, V.P., Factors responsible for multiyear changes of the ice distribution in the North Atlantic. Trudy gos. okeanograf. inst., 1969, No. 96, pp. 39-55 (12B99).
56. Tsurikov, V.L., On the formation of the ion composition and salinity of sea ice. Okeanologiya, **V**, No. 3, pp. 463-472, 1965.
57. Tsurikov, V.L., and M.I. Tsererina, Review of sea ice research in other countries. Trudy gos. okeanograf. inst., No. 76, pp. 127-212, 1964.
58. Cherepanov, N.V., New circular ice corer. In: Probl. Arktiki i Antarktiki, No. 32, Gidrometeoizdat Press, Leningrad, 1969, pp. 122-125 (1B34).
59. Shesteperov, I.A., The testing of V.L. Tsurikov's formula against observational data. Okeanologiya, **IX**, No. 4, pp. 616-617, 1969 (3B108).
60. Shesterikov, N.P., Complete freezing of polar ice. In: Probl. Arktiki i Antarktiki, No. 33, Gidrometeoizdat Press, Leningrad, pp. 39-45 (11B125).
61. Addison, J.R., Electrical properties of saline ice. J. Appl. Phys., **40**, No. 8, pp. 3105-3114, 1969 (2B535).
62. Addison, J.R., Electrical relaxation in saline ice. *Ibid.*, **41**, No. 1, pp. 54-63, 1969 (8B128).
63. Barnes, J.C., D.T. Chang, and J.H. Willand, Satellite infrared observation of Arctic sea ice. AIAA Paper, No. '301, 1970, 6 pp. (11B43).
64. Bilello, M.A., and R.E. Bates, Ice thickness observations. North American Arctic and Subarctic 1964-1965, 1965-1966. Pt. IV. Cold Region. Res. and Engng. Labor., U.S. Army, Hanover, N.H., 1969, 134 pp. (11B372).
65. Brown, J.H., Energy exchange at a sea ice boundary. Solar Energy, **12**, No. 4, pp. 469-490, 1969 (8B103).
66. Dolber, S.R., and R.T. Getman, A round-the-world-cruise by the *South Wind*. Antarctic J. U. S., **4**, No. 6, pp. 294-300 (9B178).
67. Dunbar, M., The geographical position of the North Water. Arctic, **22**, No. 4, pp. 434-441, 1969 (11B135).
68. Foster, T.D., Experiments on haline convection inducted by freezing of sea water. J. Geophys. Res., 74, No. 28, pp. 6967-6974, 1969 (10B51).
69. Frankenstein, G.E., Ice operations. USAF Nucl. Safety, 65/2, No. 1, pp. 82-83, 1970 (10B406).

70. Fristrup, B., Ice investigations, *Ibid.*, pp. 84-86 (10B189).
71. Gallery, P.D., Studying the ocean from space. Ocean Industry, 4, No. 9, pp. 31-34, 1969 (4B15).
72. Ingram, R.G., O.M. Johannesson, and E.R. Pounder, Pilot study of ice drift in the Gulf of St. Lawrence. J. Geophys. Res., 74, No. 23, pp. 5453-5459, 1969 (7B164).
73. Knapp, W.W., A satellite study of the ice in Antarctic coastal waters. Antarctic J. U. S., **4**, No. 5, pp. 222-223, 1969 (4B174).
74. Kutschale, H., Arctic hydroacoustics. Arctic, **22**, No. 3, pp. 246-264, 1969 (9B89).
75. Lake, R.A., and E.L. Lewis, Salt rejection by sea ice during growth. J. Geophys. Res., **75**, No. 3, pp. 583-597, 1970 (11B116).
76. Langleben, M.P., Attenuation of sound in sea ice. 10-500 kHz. J. Glaciology, 8, No. 54, pp. 399-406, 1969 (5B85).
77. Langleben, M.P., Albedo and degree of puddling of a melting cover of sea ice. *Ibid.*, 8, No. 54, pp. 407–412 (5B78).
78. Lock, G.S.H., On the use of asymptotic solutions to plane ice-water problems. J. Glaciology, 8, No. 53, pp. 285-300 (1B392).
79. Lock, G.S.H., J.R., Gunderson, D. Quon, and J.K. Donnely, A study of one-dimensional ice formation with particular reference to periodic growth and decay. Internat. J. Heat and Mass Transfer., **12**, No. 11, pp. 1343-1352, 1969 (6B368).
80. Lofgren, G., and W. F. Weeks, Effect of growth parameters on substructure spacing in NaCl ice crystals. Cold Region Res. and Engng. Labor., U.S. Army, Hanover, N.H. 1969, 24 pp. Ref. U.S. Govt. Res. and Develop. Repts, **69**, No. 14, p. 74, 1969 (6B163).
81. Malmberg, S.A., Astand sjavar milli Islands og Jan Mayen of hafis vid Island. Aegir, 1969, No. 22, pp. 410-411, 413 (8B147).
82. Malmberg, S.A., Hafis fyrir Vestfjördum i lok april 1969 og straummaelink i Lataröst. Natturufredungurinn, **39**, No. 3-4, pp. 234-250, 1969 (11B136).
83. Marshall, N., The icefields around Iceland in spring of 1968. Weather, 23, No. 9, pp. 368-376, 1961 (8B147).
84. Matlock, H., W.P. Dawkins, and J.J. Panak, A model for prediction of ice-structure interaction. 1st Annual Offshore Technol. Conf., Houston, Tex., 1969, Preprints, Dallas, Tex., **1**, pp. 687-694 (10B184).
85. Mosbech, P.E., Some experiments with the use of helicopter for spacial survey jobs. Internat. Hydrogr. Rev., **46**, No. 2, pp. 33-35, 1969 (2B536).
86. Muller, E.H., Sea-ice symposium in Reykjavik, Geogr. Res., **59**, No. 3, pp. 445-448, 1969 (2B105).

87. Ozawa, K., and T. Isouchi, Observations of icebergs and pack ice. J. Tokyo Univ. Fish., 9, No. 1, pp. 49-55, 1968 (5B144).
88. Paige, R.A., Stalactite growth beneath sea ice. Science, 167, No. 3915, pp. 171-172, 1970 (10B130).
89. Pounder, E.R., and M.P. Langleben, Acoustic attenuation in sea ice. Arctic Inst. N. Amer., Washington, D.C., 1967, 12 pp. Ref. U.S. Govt. Res. and Develop. Repts, 1969, No. 17, p. 80 (6B148).
90. Ross, B., Perforation of the sea-ice cover by projectile impact. J. Hydronaut., 3, No. 3, pp. 115-121, 1969 (3B42).
91. Ross, B., Penetration and fracture of sea ice due to impact loading. Physics of Snow and Ice. I, Pt. I. pp. 499-521, Hokkaido Inst. Low Temper. Sci., 1967.
92. Rouse, J.W., Jr., Arctic ice-type identification by radar. Proc. IEEE, 57, No. 4, pp. 605-611, 1969 (2B38).
93. Rouse, J.W., Jr., Arctic ice-type identification by radar. Arctic Inst. N. Amer., Washington, D.C., 1968, 34 pp. Ref. U.S. Govt. Res. and Develop. Repts, 69, No. 17, p. 80, 1969 (6B80).
94. Schmidt, E. Über die Ursache von Rissn in der Eiskappe des nordlichen Polarmeeres. Arch. Meteorol. Geophys. u. Bioklimatologie, A 18, No. 3-4, pp. 391-394, 1969 (6B161).
95. Sherman, T.D., The economics of shipping in Polar Seas. Polar Record, 14, No. 91, 479-487, 1969.
96. Sissala, J., Observations of Antarctic Ocean tabular iceberg from the Nimbus II satellite. Nature (London), 224, No. 5226, pp. 1285-1287, 1969 (9B158).
97. Stehling, K.R., Remote sensing of the oceans. Astronaut. and Aeronaut., 7, No. 5, pp. 62-67, 1969 (12B22).
98. Twiss, J.R., Jr., Antarctic sub-ice observations, 15, No. 2, pp. 109-111, 1969 (5B49).
99. Untersteiner, N., Sea ice and heat budget. Arctic, 22, No. 3, pp. 195-199, 1969 (8B133).
100. Untersteiner, N., and G.A. Maykut, Arctic sea ice. Naval. Res., 22, No. 5, pp. 12-23, 1969 (1B87).
101. Valeur, H.H., Ice conditions in Greenland waters during January-July 1963. Spec. Publ. Internat. Commiss. Northwest Atlant. Fish., 1969, No. 7, Pt. 1, pp. 29-30 (5B109).
102. Vinje, T.E., Sea-ice observations in 1968. Arbok Norsk polarist. 1968, Oslow, 1970, pp. 95-100 (8B141).
103. Vinje, T.E., Some observations of the ice drift in the East Greenland Current. *Ibid.*, pp. 75-78 (8B142).
104. Weeks, W.F., and A. Assur, Ice pressure against structures. Proc. Conf. held at Laval Univ., Quebec, 10-11 Nov. 1966. Ottawa, 1968, pp. 25-78.

SEA FORECASTS

S. I. Kahn

The Soviet Hydrometeorological Service has two large research institutes that work in the area of sea forecasts; these are the Arctic and Antarctic Research Institute (AARI) in Leningrad and the Hydrometeorological Center (HC) in Moscow. Forecasting research for purposes of commercial fisheries and oceanography is also carried out by research institutes of the fisheries industry.

Several institutes in the United States, Canada, Japan, and Scandinavian countries, among others, also do research on sea forecasts. The state of the art in sea forecasting outside the Soviet Union has reached a level whereby it is now possible to produce a day in advance charts of water temperature and wind waves for the North Atlantic and the North Pacific to meet the practical needs of shipping services.

The science of sea forecasts is developing in step with advances in physical, statistical, and hydrodynamic techniques, employing chiefly numerical methods that require the use of electronic computers.

The combination of physical and statistical techniques is based on interpretation of actual data on behavior of the sea system. Such data are used for defining the characteristics that would predict the future sea state, and the relationship between predicting characteristics and predicted values is determined by statistical methods.

The theoretical prediction of the future state of the hydrosphere on the basis of equations of hydrodynamics and thermodynamics requires the analysis of the corresponding unsteady-state processes in the ocean and the development of efficient methods of their computation.

The basis for sea forecasts is the analysis of interaction between the ocean and the atmosphere, which apparently involves the cycle ocean–atmosphere–continent. The difficulty in computing the results of these basic processes resides in the lack of accurate description of the direct and feedback relationships between the atmospheric processes and the state of the underlying surface [57].

We shall not review the broad problem of air–sea interaction. However, we deem it essential to summarize Duvanin's paper [20], which postulates a mechanism of interaction between those macroprocesses in the ocean and the atmosphere that transport heat from the equator to the higher latitudes in both hemispheres and analyzes the fluctuations of warm and cold currents and the related changes of water and air temperatures. In particular, Duvanin's comparison of the level of the Baltic Sea with air temperature over Iceland for a period of over one hundred years revealed fluctuations with a period of 5 and 2.5 years, the 5-year fluctuations being dominant.

Descriptions of large-scale processes form the basis of the long-term sea forecasts. For example, long-term ice forecasts are based on the observed fact of continuity of synoptic processes from season to season, the most distinct continuity being that between the pre-winter and winter time and the stability of the thermal anomalies in the ocean. The long-term forecasts computed on the basis of assumed mechanisms of air–sea interaction then themselves become a convenient test of the assumed interaction mechanisms.

Short-term forecasts are based on mechanisms of processes that are small-scale in time and space and permit to trace the actual relationships of different phenomena in the sea and atmosphere.

Sea forecasts predict thermal and dynamic conditions in the ocean. Forecasts of thermal conditions include those of water temperature (at the surface and various depths) and of ice (time of its appearance and disappearance, thickness, and quantity), while the dynamic conditions that are forecast are sea currents, levels, and state of the sea. Sea-state forecasts have improved greatly in recent years under the pressure of the practical need to guide ships on recommended courses. Sea states and currents are predicted merely for short periods of time, whereas the other forecasts involve a great variety of periods.

Modern sea forecasting is distinguished by the employment, as predictors, of data on the distributions of hydrometeorological variables over very large fields [28]. These fields must encompass large areas of land (including inland and marginal seas) and of the ocean so as to take into consideration all the basic thermodynamic factors that may affect future states. The fields so chosen generally range over some constant centers of action of the atmosphere such as the Siberian and the Azores highs and the Iceland low, as well as several seas. For example, in long-term ice forecasts for all the seas in the mean latitudes in the USSR one uses only two fields, namely one for the seas of the European part of the USSR and the other for those of the Far Eastern part. This approach is sufficiently broad and reduces the computational labor. The following variables are utilized in forecasting: the ground air pressure and temperature (usually in anomalies), as well as pressure at altitudes sufficient to characterize the average state of the troposphere and the water temperature of the sea. The meteorological fields are plotted directly from charts, and the values of a given variable are read off at the nodes of some grid, in most cases one with dimensions of 10° along the latitude and 5° along the longitude. The most common grids in sea forecasting are those of 63 (7×9) and of 99 nodes (9×11). The shortness of the series of data on the sea and their too low frequency force one to employ a great variety of methods for the interpretation of water-temperature data. For this purpose one can use information from multi-day stations and weather ships for specific areas, such as the Kuroshio and its counter-current, or from secular sections such as the Kola meridian, or information from averaged charts (in most cases, monthly averages).

Thus, the accuracy of sea forecasting depends heavily on the methods of collecting observations at sea, on the availability of data from an observational network of permanent unattended buoy stations, particularly on test ranges. The future of forecasting is intimately tied to deployment of a satellite network, but satellite data are already being employed for sea forecasting, even if not on routine basis.

For analytical representation of fields, expansion into infinite series of a specific form is used [13], expansion into Chebyshev polynomials being particularly widely employed. Its advantages are rapid convergence of the series, the relative simplicity of calculations, and the fact that individual terms of such series, especially the first ones, describe simple atmospheric processes such as meridional or latitudinal air transport, the convergence and divergence of air streams, etc. However, this technique is somewhat formalistic, since the physical interpretation of the terms mentioned above does not always hold, especially at high degree of expansion. To avoid this drawback one can represent the field in terms of its so-called natural components, which are calculated for this purpose from characteristic real situations. Both of these techniques of mathematical expansion are used in all methods of sea forecasting. The coefficients of the expansion series are then related to a given predicted variable by the methods of regression analysis. In many cases the predicted variable is represented analytically in the form of a series. Most of the forecasting computation work is done by computers, especially series expansion and regression analysis. Practically speaking, the number of terms in the series is limited only by the number of nodes on the chosen grid, while the number of variables being correlated is limited by the length of the observational series used in a given case. However, while experience shows that statistical relationships based on a relatively small number of arguments appear to be more stable in time, there is a need for defining methods for chosing the optimal number of variables. The widespread use of statistical methods must not, however, be allowed to supersede the unchanging principle of forecasting, namely that such forecasting must ultimately be based on the physical analysis of the mechanisms of the underlying processes.

FORECASTS OF SEA CURRENTS

A local, 1 to 5 days in advance forecast of nonperiodic currents at depths of 25 to 1200 m may be obtained from data on current velocities and the coefficients of expansion of the pressure field into its natural components [48]. Equations for the prediction of current direction and velocity have been obtained for several specific areas of the northwestern Pacific, just as it had been done earlier for the Atlantic. The analysis of these equations (including the examination of extreme cases) has shown that the change of velocity at depths of 25 to 50 and 100 to 500 m, caused by corresponding change in the pressure field, may be 1.5 times greater than the contribution due to inertia. In this case, the probability that errors will not be greater than standard deviations is 0.8 to 0.9.

One can also use the simple, practical method of calculating wind-driven currents from the atmospheric pressure gradient. There exists a nomogram [34] for this purpose, which was plotted with allowance for the wind factor. Such a nomogram permits one to perform routine calculations of wind-driven currents for different areas of the ocean 5 to 6 times faster than heretofore.

James [64] developed a plot for determining the drift current velocity from the wind velocity, duration, and fetch. Curves for calculating the velocity of the geostrophic flow from the gradient of water temperature have already been plotted for individual sea areas, such as, for example, the Gulf Stream and the Sargasso Sea.

Several papers [4, 5, 46, 65] on various problems of hydrodynamics indirectly touch on problems of current forecasting. Thus paper [65] deals with Indian Ocean's response to the changes in the southwesterly monsoon. Lineykin [36] reviews the hydrodynamic theory of sea currents. And, finally, Hackert [62] describes the pattern of currents in the Baltic Sea and gives a forecast of the changes expected in the long-term future, based on the statistical characteristics of various current variables and wind-velocity components.

FORECASTS OF SEA STATE

A summary of the state of this problem, describing the theory and the calculations of components wind-driven waves, was written by Titov [54].

In many cases, wave calculations are done for purposes of specific hydroengineering works, as in the calculation of wind parameters [67], in which the reliability of the calculated wave heights is discussed, and recommendations for the determination of heights and wave periods from the desired depth of water near the engineering works are given.

The Board of Caisson Breakwater Construction of the port of Onahama, Japan, analyzed the requirement on the various types of sea-state forecasts, depending on the type of work being undertaken, with particular attention to the definition and meaning of a given forecast [26]. The paper is concerned with the reliability of the monthly, weekly, and 3-day forecasts of sea state in terms of the purposes of the construction project as a whole and of its separate subdivisions.

Kikuchi [30] analyzes the wave heights and their changes in the swell produced by typhoons on the basis of observations in areas close to Japan. Kikuchi compares the characteristic heights of the swell-wave systems in individual areas with the typhoon's position. It was noticed that upon the typhoon's approach, the wave height increases to a maximum in one to two days; after its passage, the wave height drops in one day to a negligible value. Kikuchi gives a graphical method for predicting the maximum wave height from the pressure measured at the center of the typhoon and the distance from the typhoon center to the point for which the wave height is being calculated.

The analysis of wind-driven waves in terms of velocity of the driving wind is now routine. Such analysis permitted the development of curves relating wave height, wave length, wave period, and wave velocity with the wind velocity on the Romanian coast of the Black Sea [69].

It ought to be noted that the number of papers on sea-state forecasts decreased greatly in 1970. But, at the same time, merchant fleets and fishing ships were provided, on an increasing scale, with efficient information on the best recommended courses of navigation in the open ocean. These recommendations are developed by plotting and analysis of charts reflecting the actual sea state and by preparing charts of predicted sea states. Even though this type of forecast is undoubtedly of great practical significance and of great interest, little has been published on it. To fill this gap, we now have a manual on route navigation [1].

The choice of the optimum ship route involves the calculation of the path the ship can be expected to cover, given the speed loss that can be projected on the basis of available 3-day forecasts of the sea state. The recommendation for the ship's path for the period beyond the 3-day forecast is based on climatic data and long-term weather forecasts. The overall path thus chosen is corrected daily by means of on-board observations and short-term weather forecasts. When the ship is faced with unfavorable weather (a hurricane, or a zone of severe sea disturbance), an immediate recommendation for changing the course is issued. Thus, ships at sea are under constant control from shore, from the moment of their departure to the moment of arrival to the port of destination. The methods of calculating recommended courses are either numerical or graphical. The numerical methods employ the equations of the ship's motion with allowance for the sea state and currents. The sea-state and current forecasts are plotted on a coordinate network and then interpolated for any given point. The sea-state field is assumed invariant for periods of 6, 12, or 24 hours, depending upon the frequency of compilation of the forecast chart. The current field is also assumed to be invariant. The calculated shortest distance (along a great circle) between the initial and target points of sailing is divided into segments equal to the distance covered by the ship in a day. Cones of rays, within the limits of which the ship's optimum daily path is sought, are plotted from the ends of segments. The plotting is done by means of a special computer program, which calculates the minimal time that the ship needs to cover the given distance and gives the coordinates of the optimal path.

FORECASTS OF SEA-LEVEL FLUCTUATIONS

Major theoretical and laboratory studies have been carried out in recent years on catastrophic sea-level surges. Analysis of specific cases of such surges, as well as a search of empirical relationships useful for forecasting catastrophic sea-level variations is continuing. The results of a now disbanded research group that worked on storm surges for 19 years (until 1968) were evaluated in [23]. This group was attached to the Committee on Studies of the Coast of the North and Baltic Seas, and explored the origin and effect of storm wave surges, as well as the application of the resulting data in planning engineering works designed to protect the northern coast of West Germany from such catastrophic surges. Apart from traditional methods, this group suggested some new prediction methods, based on a numerical solution of the hydrodynamical equations, as well as some practical recommendations for planning sea dams.

The advances in the study of storm tides in the White and Baltic Seas are described in Potanin's paper [45]. Some relationships useful for forecasting water level in Arkhangelsk at high and low tides, with allowance for the tidal characteristics and predicted wind and atmospheric pressure were suggested in [44]. The forecasts are given 12 hours in advance, and the wind component is predicted along the axis of the Dvina Bay. A computer program for calculating wind fields on the basis of ground-level pressure, used for the calculation of storm tides, was developed for the White Sea area [19] but is applicable to other sea coasts. An attempt has been made to predict nonperiodic sea-level fluctuations [3], represented as a sum of a nonperiodic and a tidal component, in terms of the correlation theory of random, steady-state processes.

The analysis of hourly tide observations on the Pacific and the Sea of Okhotsk coasts of northern Japan [25] showed that the narrowness of the coastal shallow-water zone impedes the development of storm surges. Thus in 10 years there was only one station at which water level exceeded the calculated (from tidal harmonics) tide by 40 cm. This occurred 26 times at that station, with the maximum excess reaching 70 cm. The increase in sea level was caused by a combined effect of the hydrostatic pressure in the center of the cyclones and the slope of the water surface produced by the development of coastal currents, generated by northerly and northeasterly winds.

Over a period of many years, research on the problem of floods and flood prediction has continued in Leningrad. The established opinion is that floods in this area are due to the development of a long single wave. Mikhailov [39] analyzed the data of sea-level observations made of several points in the Baltic Sea, as well as data on currents in the northern part of this sea during the flood of October 18, 1967, and found that this opinion is supported by the recent data. During the October 1967 flood there were uniform currents throughout the entire water column in all the phases of the sea level's increase and decrease; there was also coincidence of maximum current and maximum sea-level phases. Data analysis showed that meteorological variables were the dominant factors in the flood. Numerical experiments on the calculation of hypothetical catastrophic level rises were also run.

The analysis of the Gulf of Finland level rises, observed in 1953-1960, showed that they fit the curve relating the gradual increase of the long wave and the change of the westerly wind-velocity component. This wave was plotted from the data for earlier years, beginning with 1903 [55].

The lows that cause storm tides on the southwestern coast of Holland produce free long waves propagating with a velocity of about 34 km/h at a sea depth of 30 m. Data for the period 1898-1966 show that such lows travel along paths lying within fairly narrow limits and that the sea-level increments are due to irregularities of the air flow caused by the passage of a front at a velocity of 31 to

36 km/h. At these velocities a resonance with the free wave is produced due to the difference between the atmospheric pressure and wind stresses on the two sides of the front. This situation causes incremental level rises [68] of 5 to 6 dm.

A method for the prediction of the sea level on the southeastern coast of the Baltic Sea [24], based on correlation of the atmospheric pressure field with the tendency of the water level to persist (inertia), enables one to forecast sea-level variations 1 to 3 days in advance. The pressure field may be analytically represented as a series expansion in natural components, calculated from typical pressure situations. Various equations yield forecasts for the warm and cold seasons of the year under the assumption of a stable low tide.

The depth, bottom topography, and shore configuration are important in tide generation in shallow-water seas. Ice and river runoff are also factors in tide phenomena. For example, fast ice decreases level variations and dense, concentrated drifting ice reduces the energy of tidal fluctuations, but, by contrast at river bars, may cause level rises due to the shallow-water effect. Quantitative relationships between level variations and individual hydrometeorological factors were derived for the shallow seas of the Arctic [41, 42]. It was found that in a rectangular bay of constant depth, a combination of tidal and surge waves may entail catastrophic level rises. Assuming that this involves a linear superpositioning of the two waves, one can analyze the surge effect separately. A problem of this type was solved [66] under unsteady-state conditions, with bottom friction and Coriolis force being ignored. The result was expressions used for preliminary estimation of tidal phenomena in the Bay of Fundy at a wind velocity of about 23 m/sec. In this case the maximum surge level was 1.2 m.

Purely statistical methods of level calculations are also being developed. Thus statistical theory was used in the determination of the maximum sea levels, that is, levels defined as those which would be exceeded only with a very small probability [12]. Most of this paper deals with the problem of whether the observational series are representative. This research yielded a regional dimensionless function of the distribution of maximum levels. When applied to the data for the Sea of Azov, this distribution fits well the first limiting distribution of maxima.

Forecasts drawn up a month in advance are intermediate-term, between storm-surge and tidal-level forecasts and seasonal predictions. In the Caspian Sea month-to-month tidal variations are important in sea-level changes [52]. Quantitative relationships between the ground-level atmospheric pressure field, represented as a series expansion in Chebyshev polynomials, and sea-level expended in natural components, have been obtained for purposes of monthly sea-level forecasts.

Long-term observations in the Caspian Sea showed that the principal cause of sea-level fluctuations is the seasonal variation in water balance, primarily river runoff [50]. The effect of seasonal density changes on sea level is insignificant. The shape of yearly curve of sea level is invariant for all points of the sea. The specific seasonal wind variations and configuration of the coastline introduce some differences in the level fluctuations at the western and eastern coasts.

There exists a statistical description of seasonal variation of sea level, with allowance for the main causative factors, for the Aral Sea [37]. The variability of the main characteristics is fairly large, especially during the colder half of the year. The data permit one to forecast with some confidence the curve of seasonal sea-level changes as a function of river runoff.

The total area that will be covered by water of the Caspian Sea (including all its bays) by the year of 2000 was predicted from historical and contemporary data [32].

FORECAST OF WATER TEMPERATURE

The differences in heat distribution during the warm and cold seasons of the year have led to different analyses and forecasts of sea-water temperature in different seasons. During the warm season there exist large vertical and horizontal temperature gradients, and one must take into account many factors that affect temperature distribution. Vertical distribution may change substantially over very small time intervals. In the colder periods, when the temperature distribution in the upper water layer is almost constant with depth, the calculations need to take account only of convection.

The techniques of short-term water temperature forecasts were summarized in [14]. Predictions for the warm season are based on the heat budget (for estimates of heat flows across the surface), the atmospheric pressure field represented analytically as an indirect characteristic of advection, and the initial temperature field similarly represented. The prediction equation is derived by the method of multiple correlation.

Owing to the complexity of processes involved in generating a given temperature field, it is recommended [64] that the effect of each factor be evaluated separately for each season, namely:

a) for the period during which the retention of the preceding state appears to be the best forecast;

b) for the period in which one must take into account heat exchange, as well as mixing and advective changes;

c) for the period for which climate values appear as the optimum approximations.

Hence, one must evaluate the initial conditions, calculate the advection and determine the heat flow through the sea surface. The next step is to determine the depth of mixing depending on calculated heat budget.

There now exist practical methods for approximate calculation of heat flow through sea surface from the data of five, four, or three observed variables, depending on the amount of information these carry [56]. The water temperature itself is represented as an expansion of the curve (in the vertical direction) and fields at the surface. Thus eigenvectors of the water-temperature field, reflecting the overall thermal conditions at the sea surface, were plotted for case of the Baltic Sea [10], and it was found that an advective transport from the southwestern part of the Sea is the most pronounced variable.

Direct observations in the shelf area of southwestern Africa [58] were used to check out the theoretically derived relationships between water temperature and several meteorological variables. It was found that air temperature is the best predicting variable, with the atmospheric pressure field and the meridional component of the wind being less useful.

Forecasting relationships between water and air temperatures over the cold season were compiled for the shallow-water (depths to 1.5 m) northern zone of the Caspian Sea. The water temperatures are the input data and are predicted one day in advance [7].

Of the work on forecasts of water temperature, let us mention paper [15] that outlines the possibilities for forecasting the surface-water temperature in the North Atlantic. Studies done for this purpose yielded estimates of the natural components of pressure anomalies in the northern hemisphere, these being estimated separately for the Atlantic Ocean, Europe, and Asia. Natural components of water temperature anomalies in the surface areas of the North Atlantic are based on data collected by 9 weather ships. Expansion coefficients for the pressure field that prevails in January were used for predicting expansion coefficients for the August water-temperature field. Another dynamic statistical method [49] for predicting macroprocesses, employing the theory of probabilistic processes, is based on observations of surface-water temperatures in the North Atlantic (Smed's squares). The time range of forecasts in this method varies; however, the best results are obtained with those for a period of three months.

The quality of the water-temperature forecast for the colder season strongly depends upon the input information. The use of data from airborne infrared surveys now permits quasisynchronous water-temperature determination over a very large sea area [38]. Heat losses over this season are calculated from simplified heat-balance equations that are based on the difference between water and air temperatures. The depth of convective mixing is precalculated from available data on the decrease of water temperature, using observations collected in preceding years. Advection-induced changes in water temperature are approximated from long-term observations, but this leads to errors in the calculation of convection in areas of fronts and meanders. This method of water-temperature prediction is now used in practice, but its time range is limited by the predictability of air temperature. In seas with permanent currents, one can use the so-called "catch-up time" method. This method was employed in the analysis of water temperatures in the Norwegian Sea and at the meridian of the Kola Peninsula [20]. Regression equations for an approximate forecast of winter temperatures on the Kola meridian, based on summer observations in the Norwegian Sea, have been derived. The effect of heat-balance variations at the sea surface upon the changes of temperature of the water in an area extending from the Faroe–Shetland Trench to the Kola meridian was traced from 11-year observations.

A study of even larger (in terms of time intervals), large-scale processes in the ocean and atmosphere is continuing. It was shown that there exists a correspondence between the hydrological variables in the Northern Pacific and the climate of North America [63]. This study revealed a close relationship between average air and sea-surface temperatures, which exhibits a periodicity of two and four to five years. It was shown that extreme deviations (from the average) of water temperature in the ocean are synchronous with large-scale changes of meteorological conditions in Peru, Japan, and in California. It was also shown that there exists an inverse relationship between the hydrological processes occurring in the North Pacific and the North Atlantic.

Smirnova [52] describes the periodicity of changes of the water temperature in the Bering and Norwegian Seas. She found periodicities of 4.7, 11, and 18 years. She also proved the existence of a relationship between the average temperature in July at the Kola Peninsula meridian and the long-term fluctuations of the potential of the tide-generating force, as well as between the 11-year solar-activity cycle and oscillations of the terrestrial pole.

Sarukhanian and Smirnov [47] introduced a "genetic" method of long-term prediction of water temperature based on the same input data as Smirnova's [52] technique. They were able to predict the temperature in the Kola meridian area in an anomalously cold year with much greater reliability than Smirnova.

Several 1970 publications deal with application of temperature forecasts to the needs of the fisheries industry. Some of these papers are qualitative, i.e., do not present any quantitative relationships but merely indicate the overall conditions in each area of the ocean that are conducive to appearance of fish schools. There are also, however, some purely predictive studies in which the atmospheric pressure, represented as expansion into Chebyshev series, is related to fluctuations of water temperature known to control the catch of individual fish species (11). Strictly speaking, the atmospheric processes are not directly related to fish catches, but to sea conditions. It is the latter that determine the size of the different fish schools, and thus the size of the ensuing catches. This chain of interdependences is not quite evident in the above-cited works, but the practical results proved satisfactory. There have been attempts to solve the same practical problem by making use of atmospheric-pressure data in some other ways, namely, by analysis of isoanomaly charts, of the centers of atmospheric action, etc. [59].

Time and space changes in water temperatures are analyzed in terms of the sizes of lobster catches in the coastal area extending from Newfoundland to New York with a view toward developing long-time forecasts [61].

The bottom concentrations of pelagic fish species are analyzed in connection with seasonal changes of water temperature [6].

The biological productivity of waters of northwestern Africa is controlled by the effects of the monsoon. This is confirmed by water dynamics [33]. In winter, the northwesterly wind drives the surface waters away from the coast and thus intensifies the upwelling of water from the depths, while in summer a hydrologic front is formed. The French tuna fleet finds the best fishing near the coast waters of Africa [60] because of the effect of the monsoon. The water temperature in the saira catch area during the period preceding the beginning of the fish's approach is predicted by a synoptic analysis and depends on the structure of the thermobaric field [22]. The calculation of the cyclone-frequency indexes and the study of features of the atmospheric-pressure anomalies are closely involved with these practical needs.

FORECASTS OF ICE PHENOMENA

Forecasting ice phenomena involves several different problems that are related to thermal and dynamic processes in the sea. The number of variables that must be predicted is very high. They include the onset times of the main icing phases (the appearance of ice, fast ice formation, its breakup and disappearance, the appearance of ice of any given concentration), the thickness of ice and its transition through certain grades, concentration and drift, ice field coverage, and the position of the ice edge in the open sea. At present, ice forecasting involves prediction of individual characteristics of drifting and fast ice, and development of methods for short-, long- and very long-term predictions. Ice forecasts deal with autumn–winter processes that control the appearance and the growth of ice, as well as with spring events which are responsible for ice decay, melting, and disappearance.

The needs of year-round shipping in seas of midlatitudes have produced a great demand for forecasts for the open-sea areas over the entire winter season, in contrast to the previous situation, where the predictions needed were the ice-phase forecasts for the principal port areas. Forecasting of the position and size of ice clusters, hummocking, etc., in the Arctic is a routine task of great importance [35]. Ice forecasting tends to be a science oriented toward practical applications, theoretical studies being fairly scarce and dealing chiefly with ice drift. Such theoretical studies include investigations of the effect of turbulence on the average velocity of ice drift [53], and the calculation of ice drift and concentration [19]. The drift of ice floes whose area is much greater than their thickness was also explored [21]. In this case the equations used in calculations of ice movement were those for the unsteady state and allowed for the Earth's rotation, friction at air–ice and ice–water boundaries, as well as for ice floe interaction (by means of Laykhtman's theory). The calculated results agree with observations [21].

The possibility of forecasting ice drift is tested on data obtained in special observations in the White Sea [40]. The analysis of these data yielded the wind factors for drift in various areas, the drift characteristic in the coastal zone, and the magnitude of ice discharge through straits.

The results of ice-drift observations in the East Siberian Sea [16] revealed a connection between various types of ice distribution and large-scale atmospheric processes.

A 1970 book dealing with ice navigation and the scientific support for it [51] has a chapter on forecasting of ice conditions. It reports that in the United States and Canada ice forecasts are based on the analysis of the ice balance of the Arctic Ocean and adjoining basins, which is treated as a part of the heat balance of the planet. The studies of Soviet scientists (N. N. Zubov, V. V. Shuleikin, etc.) are widely used in America for the purpose of forecasting freezing, thickness, and redistribution of ice. Attention is also paid to the relationship between ice and meteorological forecasts, and the reliability of ice predictions. The basic methods used for prediction are the method of analogs, statistical comparisons, and the elucidation of underlying physical patterns.

Special manuals on preparation of ice forecasts to meet the needs of icebreakers and cargo ships are being developed. Much attention is also devoted to ice forecasting manuals designed to meet the needs of submarine navigators.

The ice-forecast manuals compiled for the Baltic Sea are concerned primarily with the prediction of the time of onset of ice and the achievement of various ice thicknesses.

Japanese scientists are also engaged in the study of ice-drift forecasts and the physicochemical properties of ice.

Data for atmospheric processes over very large areas are widely used in ice forecasting. Almost all long-range ice forecasts are based on such observations. As in the case of water-temperature predictions, both the controlling factors and the predicted variable itself (for example, the concentration of ice floes) may be represented analytically as a series expansion. An attempt has been made to predict the dates of ice appearance and freedom from ice in the Seas of Japan and Okhotsk by means of series expansion, employing as data the preceding (1 to 3 months) distributions of temperature and air pressure [27].

Some statistical characteristics of ice coverage are described by Karakash [29], who develops a previously suggested hypothesis on the possibility of predicting the thermal (including ice) conditions of a forthcoming winter from the data on the pressure and temperature fields of the previous winter, bearing in mind that it is during the cold season of the year that the sea–air interaction is most pronounced. Karakash also makes use of the fact that annual and semiannual fluctuations of hydrometeorological variables tend to be stable, and derives an equation for prediction of ice coverage and several other characteristics of ice in seas of midlatitudes, with a forecasting time range of up to 12 months.

A relationship between the pressure field in winter and subsequent conditions in the Arctic seas was derived [43]. It was found that a stable cyclonic or anticyclonic circulation in the atmosphere has a pronounced effect on the ice movement in summer and on several other phenomena.

The periodogram analysis of atmospheric pressure fields over the Arctic revealed 14-month periods of fluctuations of pressure gradients and corresponding anomalies of air transport. In connection with this it has been shown [17] that such periodic phenomena ought to cause 6- to 7-year fluctuations in ice coverage, with an east-to-west shift of fluctuations. Correlation functions between the ice coverage in some seas and also between ice coverage and the coordinates of the terrestrial pole tend to support this theory.

In some cases there exists a need for predictions of the thickness of the drift and fast ice. The methods for such predictions tend to be based on assumed relationships between ice and changes of hydrometeorological parameters of very large atmospheric and hydrologic fields. However, fairly simple methods are also used, for example, in the calculation of time required for ice growth to a thickness of 20 to 25 cm, a method based on the dates of stable ice formation [18] at several polar stations.

The investigations of the Arctic ice carried out by scientists of the University of Washington [70] aim at the development of a numerical method for the calculation and forecast of ice growth and for estimating temperature changes within thick ice.

A review on sea ice in the Sea of Okhotsk [2] by Akagawa covers the basic theoretical and empirical results of ice studies. It includes an analysis of seasonal and annual fluctuations of ice cluster areas (with a 6- to 7-year periodicity) and throws light on the state of the science of forecasting.

LITERATURE CITED

The number in parentheses refers to the number of the citation in the Soviet Abstracts Journal (Referativnyy Zhurnal) Geophysics 1970.

1. Abuziarov, Z.K., and K.M. Sirotov, Recommended ship courses in oceans, Gidrometeorol. n. i. tsentr. SSSR, 1970, 90 pp.
2. Akagava, M., Ice of the Sea of Okhotsk. Kishiyu Kenkiyu Noto 1969, No. 101, pp. 449-479 (1970, 10B133).
3. Alekseyev, G.V., Some problems in forecasting nonperiodic sea level fluctuations with the aid of linear regression equations. Trudy Arkt. i Antarkt. n. i. inst., 1970, No. 291, pp. 68-78.
4. Altman, E.N., and D.M. Tolmazin, Method of current calculations and water exchange in the Kerch Strait. Okeanologiya, 10, No. 3, pp. 438-447, 1970.
5. Bakhmalov, F.G., Calculation of currents in the Caspian Sea (by the Ekman-Felzenbaum method) under the prevailing wind fields. In: Materialy nauch. teor. konferentsii molodykh uchenykh. Akad. Nauk Azerb. SSR. Book 3, Baku 1969, pp. 43-45 (1970, 11B60).
6. Bernikov, R.G., Seasonal changes of temperature and water salinity in the area ranging from Cape Blanc to the port of St. Louis. Trudy Atlant. i. inst. rybn. khoz. i okeanogr., 1969, No. 22, pp. 13-29 (1970, 8B164).
7. Barabash, A.K., Methods of water temperature forecasts and appearance of ice at the northern coast of the Caspian Sea. In: Sbornik Rabot Astrakhan. gidrometeorol. observ. 1970, No. 1, pp. 103-111.
8. Bochkov, Y.A., E.I. Sarukhanian, and N.P. Smirnov, Principal patterns in long-term water temperature fluctuations in the

Barents Sea and their relation to geophysical processes. Trudy polyarn. n. i. inst. i proyektn. inst. morsk. rybn. khoz. i okeanogr., 1968, No. 23, pp. 104-115.
9. Veselov, Ye. P., and B. Kh. Rybak, Calculation of wind fields for purposes of estimation of storm surges. Trudy Gidrometeorol. n. i. tsentra SSSR, 1969, No. 56, pp. 52-59 (1970, 4B113).
10. Vildanova, M. I., Approximation of temperature fields in the Baltic Sea with the aid of natural components. Nauchn. trudy Obninsk. otd. geogr. obshch. SSSR, 1968, Book 1, pt. 1 pp. 73-78 (1970, 2B65).
11. Vyalov, Yu. A., The use of atmospheric pressure data in forecasting fish catches. In: Atlanticheskiy Okean. Rybopoisk. issled. No. 2, pp. 65-73, Kaliningrad, 1969 (1970, 5B162).
12. German, V. Kh., and B. Kh. Glukhovskiy, Statistical method for the calculation of maximum sea levels. Trudy Gosud. okeanograf. inst., 1970, No. 98, pp. 117-136 (1970, 9B82).
13. Glagoleva, M. G., and L. I. Skriptunova, Methodological Letter No. 2. Derivation of correlations for the calculation and forecast of vertical water temperature distribution. Gidrometeorol. tsentr SSSR, 1964, pp. 1-68.
14. Glagoleva, M. G., and L. I. Skriptunova, Methods for short-term forecasting of sea-water temperature. *Ibid.*, 1970, 23 pp.
15. Glagoleva, M. G., On the possibility of predicting areal water temperature distribution. Trudy Gidrometeorol. n. i. tsentra SSSR, 1969, No. 51, pp. 75-85.
16. Gorbunov, Yu. A., and K. I. Chukanin, Redistribution of ice in the East Siberian Sea under the influence of atmospheric processes. In: Problemy Arktiki i Antarktiki, 1970, No. 35, pp. 95-100.
17. Gudkovich, Z. M., E. I. Sarukhanian, and N. P. Smirnov, Polar tides in the atmosphere at high latitudes and fluctuation of ice coverage in the Arctic seas. Dokl. Akad. Nauk SSSR **190**, No. 4, pp. 954-957, 1970 (1970, 7B144).
18. Gushchenkov, Ye. M., Methods for prediction of time needed to achieve ice thicknesses of 20 to 25 cm. Problemy Arktiki i Antarktiki, 1969, No. 32, pp. 45-51.
19. Doronin, Yu. P., Method for calculating ice concentration and drift. Trudy Arktich. i Antarktich. n. i. inst., 1970, No. 291, pp. 5-17 (1970, 10B67).
20. Duvanin, A. I., Model of interaction between macroprocesses in the sea and atmosphere. Okeanologia, **8**, No. 4, pp. 571-580, 1968 (1970, 1838).

21. Yegorov, K. L., Theory of drift of ice floes in a horizontally heterogeneous wind field. In: Problemy Arktiki i Antarktiki 1970, No. 34, pp. 71-78 (1970, 10B69).
22. Zavernin, Yu. P., Some aspects of atmosphere-hydrosphere interaction in the northwestern Pacific. Izv. Tikhookeansk. n. i. inst. rybn. khoz. i okeanograf., 1969, No. 68, pp. 67-78 (1970, 8B222).
23. Zavernin, Yu. P., Concluding remarks on the results obtained by the now-disbanded research group for problems of storm surges and recommendation on the utilization of these results in the construction of sea dams. Küste, 1969, No. 17, pp. 81-103 (1969, 12B142).
24. Zorina, V. A., Sea-level predictions for the southeastern Baltic Sea. Trudy Gosud. okeanograf. inst., 1970, No. 98, pp. 39-94 (1970, 10B93).
25. Isozaki, I., A study of sea level changes induced by meteorological disturbances on the coast of Japan. IV. Storm surges on the Pacific Ocean and the Sea of Okhotsk shorelines of northern Japan. Nihon Kaiyo, 25, No. 4, pp. 191-200, 1969 (1970, 7B188).
26. Yoshimura, I., The accuracy of sea-state forecasting. Kohwan, 46, No. 3, pp. 30-33, 1969 (1970, 2B166).
27. Kahn, S. I., A method for predicting autumn and spring ice phases represented analytically in the form of mathematical series (based on the examples of the Sea of Okhotsk and the Sea of Japan). Trudy Gidrometeorol. n. i. tsentra SSSR, 1969, No. 51, pp. 152-163.
28. Kahn, S. I., State of ice prediction techniques for the seas of the USSR. Okeanologiya, 7, No. 5, pp. 786-791.
29. Karakash, A. I., Ice forecasts for the mid-latitude USSR seas. Trudy gidrometeorol. n. i. tsentra SSSR, No. 56, pp. 101-119 (1970, 3B109).
30. Kikuchi, S., The height of swell waves and the relative position of typhoon near Kyushu Island. J. Meteor. Res., 21, No. 6, pp. 309-319, 1969 (1970, 6B207).
31. Kislyakov, A. G., The relationship between water temperatures in the Norwegian and the North Cape currents. Trudy Polyarnogo n. i. i proyekt. inst. morsk. rybn. khoz. i okeanograf., 1968, No. 23, pp. 143-156 (1970, 1B88).
32. Konstantinov, V. A., The state of the bays in the Caspian Sea. Izv. Akad. Nauk Azerb. SSSR, Ser. Nauk o Zemle, 1969, No. 5, pp. 69-73 (1970, 7B200).
33. Kopytov, V. I., and V. N. Yakovlev, Generation of productive zones of northwestern Africa. Trudy n. i. inst. rybn. khoz. i okeanograf., 1969, No. 22, pp. 3-12 (1970, 8B165).

34. Krasyuk, V.S., and Ye.M. Sauskan, Calculation of velocities of wind-driven currents in the ocean. Meteorol. i gidrolog., 1970, No. 9, pp. 68-74.
35. Lebedev, A.A., The probability of encountering ice in the Barents and Greenland Seas and in the North Atlantic. Trudy Polyarnogo n. i. i proyekt. inst. morsk. rybn. khoz. i oceanograf., 1968, No. 23, pp. 138-142 (1970, 1B157).
36. Lineykin, P.S., Gidrodinamika morskikh techeniy (Hydrodynamics of Ocean Currents), Obninsk, 1969, 55 pp.
37. Lvov, V.P., Seasonal changes of mean level in the Aral Sea. Trudy Gosud. okeanograf. inst., 1970, No. 98, pp. 107-116 (1970, 10B175).
38. Mileyko, G.S., IR surveys of sea surface temperature for purposes of calculations and prediction. Trudy n. i. gidrometeorol. tsentra SSSR, 1969, No. 51, pp. 86-92.
39. Mikhaylov, Yu.D., and K.S. Pomeranetz, Hydrometeorologic conditions during the flood of Oct. 18, 1967 in Leningrad and its numerical calculation. Trudy Gosud. gidrometeorol. inst., 1969, No. 173, pp. 110-118 (1970, 5B113).
40. Moiseyev, V.S., On the possibility of ice drift prediction. In: Sbornik rabot Arkhangelsk. gidrometeorol.observ., 1970, No. 7, pp. 32-88 (1970, 9B113).
41. Mustafin, N.V., Method for prediction of nonperiodic sea level fluctuations. Problemy Arktiki i Antarktiki, 1969, No. 32, pp. 37-44 (1970, 2B62).
42. Mustafin, N.V., Factors responsible for tidal sea level fluctuations in Arctic seas. Problemy Arktiki i Antarktiki, 1970, No. 34, pp. 5-12 (1970, 10B96).
43. Nikiforov, Ye.G., Ye.N. Chaplygin, and A.O. Shpaykher, Pressure systems and dynamic processes in Arctic seas. Okeanologiya, 9, No. 5, pp. 782-790, 1969 (1970, 5B108).
44. Popov, V.P., Short-term prediction of sea level in the port of Arkhangelsk during the ice-free period. In: Sbornik rabot Arkhangelsk. gidrometeorol. observ., 1970, No. VII, pp. 62-71.
45. Potanin, V.A., The status of knowledge of storm surges in the Barents and White Seas. In: Sbornik rabot Murmansk. gidrometeorol. observ., 1969, No. 2, pp. 52-65 (1970, 9B117).
46. Sarkisyan, A.S., and A.A. Serebryakov, Examples of calculating equatorial currents. Morsk. gidrofizich. issled. 1970, No. 1 (47), pp. 60-69 (1970, 11B54).
47. Sarukhanian, E.I., and N.P. Smirnov, The use of a genetic method for the prediction of long-term water temperature fluctuation in the Barents Sea. Okeanologiya, 10, No. 4, pp. 614-622, 1970.

48. Sauskan, Ye. M., Short-term variations of currents in the northwestern Pacific and their prediction. Trudy Gidrometeorol. n. i. tsentra SSSR, 1969, No. 51, pp. 3-15 (1970, 3B145).
49. Seryakov, Ye. I., and V. V. Strela, The use of a dynamic and statistical method in long-term forecasts of sea temperature. *Ibid.*, 1969, No. 51, pp. 93-100.
50. Skriptunov, N. A., Seasonal variability of sea level in the Caspian Sea. Trudy okeanograf. inst., 1970, No. 98, pp. 95-106 (1970, 10B172).
51. Smirnov, V. I., Ice Navigation Abroad and Its Scientific Support. GMI Press, 1970, 223 pp.
52. Smirnova, K. I., Patterns in the variation of sea level along the shore of the Caspian Sea. Trudy Gidrometeorol. n. i. tsentra SSSR, 1969, No. 51, pp. 64-74 (1970, 3B153).
53. Timokhov, L. A., Relationship between turbulence-induced and mean motion of ice. Problemy Arktiki i Antarktiki, 1970, No. 34, pp. 68-70 (1970, 10B68).
54. Titov, L. F., Vetrovyye volny (Wind Waves). Gidrometeoizdat Press, Leningrad, 1969, 294 pp. (1970, 7B92K).
55. Fredzon, A. I., and A. A. Popkov, Improving the accuracy of the curve of the relationship between the gradual growth of a long wave and the change of the westerly component of wind velocity in the Gulf of Finland. In: Sbornik rabot Leningrad. and Petrozavodsk. gidrometeorol. observat., 1970, No. 6, pp. 70-75 (1970, 9B83).
56. Sheremetyevskaya, O. I., Practical methods for the evaluation of the heat balance in the North Pacific. Trudy Gidrometeorol. n. i. tsentra SSSR, 1969, No. 51, 120-131.
57. Yudin, M. I., Theory of a physical and statistical method for very long-term weather forecasts. Trudy Glav. geofizich. observ., 1968, No. 201, pp. 3-7.
58. Yakovlev, V. N., The effect of local meteorological processes on the hydrologic conditions in the shelf area of southwestern Africa. Trudy Leningrad. gidrometeorol. inst., 1969, No. 37, pp. 14-21.
59. Yakovlev, V. N., and Yu. A. Vyalov, Allowance for hydrologic anomalies in developing commercial fisheries forecasts. In: Atlanticheskiy Okean. Rybopoiskovye issled., No. 1, pp. 15-22, Kaliningrad, 1969 (1970, 5B160).
60. Doeuff, D., Les Campagnes des pêcheurs bretons dans la région de Dakar: contribution à l'étude des relations entre le milieu océanographique et les lieux de pêche. Norois, **16**, No. 64, 1969 (1970, 9B171).
61. Dow, R., Cyclic and geographic trends in sea-water temperature and the abundance of American lobster. Science, **164**, No. 3883, 1969 (1970, 3B156).

62. Hackert., Die Strömungsverhältnisse im Greifswalder Boden bei Ost- und Westwinden. Wasserwirtsch.–Wassertechn., 19, No. 6, 1969 (1970, 2B116).
63. Isaacs, J.D., The North Pacific study. J. Hydronaut., 3, No. 2, 1969 (1970, 1B137).
64. James, R.W., Ocean Thermal Structure Forecasting. Washington, 1966.
65. Lighthill, M.J., Dynamic response of the Indian Ocean to the onset of the southwest monsoon. Philos. Trans. Roy. Soc. (London), A **265**, No. 1159, 1969 (1970, 4B99).
66. Rao D., Effect of traveling disturbances on a rectangular bay of uniform depth. Arch. Meteorol. Geophys. und Bioklimatol., A 18, No. 1-2, 1969 (1970, 2B59).
67. Suarez, B.P., Ola de calculo. Lab. puertos [Publ.], 1968, No. 5-8 (1970, 6B124).
68. Timmerman, H., Variations of the sea level along the Dutch coast and some meteorological characteristics. (Abstract). Ann. Meteorol., 1969, No. 4 (1970, 4B151).
69. Trufas, V., and Selariu O., Regimul valurilor mǎrii Negre la litorǎlul romanesc. Stud. geogr. Doborgii, Bucurest, 1969 (1970, 8B115).
70. Untersteiner, N., Sea ice and heat budget. Arctic, 22, No. 3, 1969 (1970, 8B133).

TECHNOLOGY OF OCEANOGRAPHIC RESEARCH

A. D. Dobrolvolskiy, B. S. Zalogin, and A. N. Kosarev

As the interest in the ocean increases, the techniques of oceanographic research tend to change. Whereas 15 to 20 years ago most of that research was theoretical, now much of it relies on observation. The stimulus for this change was the discovery of what was first thought to be "anomalies" of currents, such as the Cromwell–Lomonosov Current, the Gulf Stream undercurrent, and the Antilles–Guiana Countercurrent.

Among the most important publications dealing with oceanological research are reviews of accomplishments and announcements of new technology and exploration activities. For instance, a review [33] describing the activities of the U.S. Naval Oceanographic Office (NAVOCEANO), established in 1962 for the purpose of conducting comprehensive research on the ocean, reports that this office now owns 17 vessels, several aircraft, automatic buoy stations, deep-water submersibles, and an array of other modern research aids. NAVOCEANO has an Airborne Oceanography Division, and all its data are processed in close collaboration with the National Oceanographic Data Center (NODC). NAVOCEANO's mission is scientific support of the U.S. Navy, in particular, the submarine fleet.

Oceanography is expected to attract an increasing interest in the U.S. from large and small enterprises and banks. The capital investment into oceanographic research is thus expected to increase. In 1968 the American Government allotted $41 million to nine oceanographic programs involving more extensive observation of the sea and weather, undersea mining in deep-water areas, construction of new vessels, peaceful utilization of the seas within the framework of international cooperation, etc. [19].

The availability of observational data on the various areas of the ocean is analyzed on the basis of Soviet and American publications, with particular reference to the overall distribution of deep-sea hydrological stations and their seasonal distribution at various depths [1]. This information is exceedingly valuable to those analyzing the processes occurring in the ocean on the basis of literature data and to those planning expeditions to collect new material.

Walden [82] reviews the capabilities of various methods and technologies of hydrometeorological observation at sea, as well as the results obtained. In particular, he makes some interesting suggestions concerning the improvement of observations with the aid of buoy stations. Thus he thinks that hydrological variables should be measured at 20 levels down to the depth of 2000 m. Between the 60th parallel north and 60th parallel south, the suggested distance between buoys is 400 miles in the nearshore zone and 500 miles in the open sea. A total of 8 meteorological and 11 oceanographic (including hydroacoustic) variables should be measured, with shortwave transmission of data to collection points as many as 2000 miles away. The buoys should be capable of self-sustained operation for periods of 9 to 12 months, and the network of 400 buoys should be supported by 10 ships. A buoy should be built to last at least 5 years.

Some interesting observations were made with the aid of devices not commonly employed in oceanographic research [79]. Thus temperature and currents were measured with special meters and sensors mounted on the Bornholm-2 mast station at a depth of 48 m. The observations lasted from mid-May to mid-October. The current meters, spaced at 6 m intervals, were mounted at depths of 12 to 42 m, with a temperature sensor mounted 2 m below each current meter. The end of June and the beginning of July proved to be the period when the short-period pulse was transformed to an inertial flow. From mid- to the end of August, the period of the motion corresponds only approximately to the inertial period, since the water flow in this period was probably induced by wind that was active in the area before the beginning of the observations.

Finally, an interesting example of synoptic oceanographic investigations, by the Edinburgh Oceanographic Laboratory, is described [35]. This laboratory has been conducting regular quasisynoptic observations in the North Sea since 1930 and in the North Atlantic since 1939, with the aid of a towed plankton recorder (TPR) at the standard depth of 10 m, followed by plankton analysis. The TPR are towed by commercial or hydrographic vessels of various countries. The results are published as atlases of seasonal and long-term variations of the distribution, mass, and composition of plankton. A submerging TPR is now being designed that will permit a simultaneous recording of several physical variables in the upper 100 m layer of the ocean.

Many publications systematically describe the research fleets and oceanographic instruments available in the various countries. Among such publications the annual reference compilations, in the form of tabulated specifications for research vessels, should be of considerable help in planning and coordinating international sea expeditions [34].

To exchange experience and thereby to help the development of marine technology, national or international symposia are conducted.

An example of such a special conference is the 1969 Brighton (Great Britain) conference which deals with the problems of utilization of oceanographic equipment and with data processing. Most papers presented at this conference reported on data collection by means of unmanned submersible or surface buoys. It was suggested that synoptic observations over large areas could be carried out by means of merchant ships towing sensor carriers capable of collecting data over a specified depth range. Most of the unmanned stations described collect data in a form suitable for direct input into computer's [76].

The proceedings of such sessions greatly contribute to development of oceanographic sciences. Special reviews of oceanographic instrumentation, published in the Soviet Union and some European socialist countries [25] serve the same purpose.

Oceanographic equipment is described not only in reference works but also in monographs which deal with the past, present, and future of oceanographic equipment. Some of these monographs evaluate the paths of development of oceanographic instrumentation.

The pace of development of technology of oceanographic research has not been uniform across the board. Thus the numerical data-collecting techniques produce more information on the processes in the sea than the available processing facilities can handle, but the potential of the submersible systems is not fully utilized by oceanographers, one reason being that the manned underwater systems were developed without due regard to either the experience previously gained or the recent achievements in human engineering.

An arrangement of underwater habitats made up of several cylindrical or spherical compartments is suggested as the optimum underwater research facility. This facility should be provided with a water-ballast system allowing independent submergence and surfacing of the individual units, as well as with a docking and transfer system [50].

The review papers are especially valuable since they provide, in a condensed form, great amounts of information on a variety of topics pertaining to the state and organization of oceanographic research throughout the world.

In recent years, oceanographic research has come to employ observational techniques of increasing variety and sophistication. This resulted in an increasing flow of information and concomitant increase in facilities for its processing. These results are discussed at great length in special papers, monographs, and reviews, abstracts of which are printed in the Oceanology sections of the Soviet abstract journals (Referativnyy Zhurnal) Geophysics and Geography.

This vast field cannot be surveyed in its entirety in a short review, nor would it be advisable to attempt it. For this reason we have confined ourselves to the most significant work which reflects the major trends in direct (cruise-type) oceanographic research. Even though this work is being carried out by a variety of scientific and commercial organizations in a variety of ways, it and the techniques employed can be classified under several common headings.

I. NEW RESEARCH VESSELS

Ships with several different functions and thus different designs have been commissioned in recent years.

The fishery research vessel *Akademik Knipovich* (displacement 3920 tons, speed 12.5 knots, range 13,000 miles) became operational in October 1966. It is designed for comprehensive oceanographic and fisheries research and survey work in the ocean. The ship has modern navigational and search equipment, gear for fishing in water and at the bottom, as well as oceanographic instruments, and carries 13 specialized laboratories [15].

The *Papa* weather station in the Pacific is now serviced by the Canadian weather ships *Vancouver* and *Cuadra*, each displacing over 5000 tons, and capable of a top speed of 18 knots. They are equipped with stabilizers and bow propellers, special radar, automatic wind-data processors that can be collected to heights of 30 km, water-temperature recorders, radiometers, etc. [61].

An unmanned navigable sailboat, described as "a support station and mobile platform" and equipped with an automatic navigation system allowing it to follow a prescribed course, maneuver at sea, and maintain its position at a given point with the aid of an electronic orientation system, is described in [71]. The sailboat can be used instead of a buoy platform for collecting and transmitting oceanographic and meterological data. It can also be used as an automatic weather-broadcasting station, and since its inherent noise is low it is also suitable as a platform for hydroacoustic measurements.

Intensive geologic and geomorphologic surveys of the continental shelves, designed to find underwater oil and gas deposits, have spurred the development of highly specialized ships, and many vessels of different designs were produced for the study of sea-bottom topography. One of these is the R/V *Komet* (displacement of 1600 tons), of the German Federal Ministry of Transportation, which carries 6 launches. Together with the mother ship, the latter allows simultaneous echo sounding of the bottom over a large area of the shelf [48].

A fundamentally new technique for bathymetric surveys of the Canadian Arctic shelf employs a Hovercraft-carried echo sounder fitted into a fish-shaped housing that can be easily lifted out of the water when the vessel travels over land or ice-covered areas. In tests, the survey rate, with continuous recording of the profile, was 25 to 30 knots, the top speed of the vessel being 60 knots. The housing of the echo sounder can also accommodate other meters and sensors [72].

The Dutch catamaran *Duplus* is 40 m long, displaces 1200 tons, and has a top speed of 8 knots. It has cigar-shaped hulls and can easily submerge to depths of 100 m. A diving bell and other heavy equipment can be lowered from the vessel, which carries a 75-ton gantry crane and a winch. A special system of four propellers stabilizes the vessel in a single plane [53].

There is some tendency to build fewer general-purpose oceanographic vessels and concentrate on more and more specialized ships. The recent trend has been toward sea-bottom research equipment, primarily vessels of the catamaran type that can perform underwater drilling and other operations on the ocean bottom.

II. UNDERWATER VEHICLES

Ocean survey with various kinds of underwater vehicles is now typical of oceanography. These vehicles differ widely in design, function, equipment, research potential, and operation. Some of these vehicles will be described below.

The *Archimèdes* bathyscaph, property of the French Navy, is used by it jointly with the National Center for Scientific Research. The electrically propelled vehicle can submerge to 11 km, carries a continuously operating resistance thermometer accurate to ±0.05°, a pressure gauge, a compass, a thermoanemometric current meter with a velocity range from 0.1 to 10 cm/sec, and has provisions for other instruments, which are mounted when needed (for instance, an underwater-sound-propagation meter, a pH meter, or a gravimeter). The data are recorded by a 12-channel digital tape recorder. On the outside of the sphere there are two cameras which yield a stereopair of the scene viewed from the porthole, and a manipulator, which is a bridge crane with a bottom grab for picking up individual objects and sediment samples as heavy as 80 kg. The bathyscaph also carries a 0.7 m corer and a drilling rig with a diamond drill for taking core samples 60 cm long and 5.5 cm in diameter. The benthos is collected with a special suction pump mounted on the manipulator; water samples are taken with 20 one-liter sampling bottles that are controlled from inside the sphere.

The bathyscaph allows visual observation of the underwater scene, observation under direct control of the researcher, accurate sampling and collection of desired specimens, checking and testing the underwater communications (cables) as well as gas and oil pipes, and determination of the distribution of physical properties over depth [37].

The mesoscaph *Ben Franklin* has positive stability since its hull compression coefficient is lower than the compressibility of water, so that the vessel can be kept at the required depth with no ballast. It carries equipment for measuring and digital recording on magnetic tape of the temperature and salinity of the water, the velocity of sound in water, the pressure, as well as systems for measuring and recording the relative speed and direction of current. It also has a meter for measuring light absorption by water and the intensity of natural underwater illumination, a gravimeter, a magnetometer, stereocameras, an acoustic seismic profiler, a meter for measuring reverberation and absorption of sound by the sea bottom, as well as a water-turbulence meter [24].

The *Deepstar*, a special U.S. vehicle for studying the boundary zone at the sea bottom, has instruments for investigating the physical and chemical properties of water and bottom sediments, namely a thermistor temperature sensor to be inserted into the bottom sediments, a water-temperature sensor, samplers for collecting water samples from the near-bottom layer, a current meter with a Savonius rotor, and a fluorescein release can. All the instruments are mounted on a cantilevered frame on the front part of the vehicle and are remotely controlled. Motion picture cameras are available for filming pelagic and bottom-dwelling organisms [64].

A similar vehicle, the *Spurv*, designed at the Applied Physics Laboratory of the University of Washington, is intended for precise and continuous recording of oceanographic and geophysical information in the near-bottom layers. This is a self-propelled vehicle, controlled from the support vessel with the help of an acoustic command- and servo system. Its top speed is 6 knots, the maximum operational depth is 3500 m, and the operational cycle is 5 hours. Onboard instrumentation measures the temperature, hydrostatic pressure, and sound velocity in the water, as well as the distance between the moving vehicle and the ocean floor. The signals from the sensors are recorded on magnetic tape in digital form, and are then processed by standard computers. The temperature sensors are a thermistor generator accurate to within ±0.0004° to 0.0009°, with a repetition rate of 12 sec^{-1}, and a crystal thermometer accurate to ±0.00056° at a repetition rate of 1 sec^{-1}. The pressure meter is a Vibrotron device accurate to ±0.0026 kg/cm^2, with a repetition rate of 12 sec^{-1}. The distance between the moving vehicle and the ocean bottom is determined by an echo sounder every second with an accuracy of 0.13 m. The sound velocity is measured by a cycling meter, recording data every 12 sec^{-1} to within ±0.013 m/sec. The sequence of operation of all the instruments is controlled by means of an accurate timing system.

In addition to the deep-water submergence vessels, there are systems for work at depths of the order of 300 m, such as the *Guppy*. A British company is developing an underwater vessel with a high-power propulsion plant for work at depths to 180 m. It will be capable of traveling at speeds of 2.5 knots both over rugged sea floor and on the surface.

The availability of the great variety of vehicles will offer the potential for a great deal of diverse underwater research. A review has been published on functional designs and possible performance of various underwater vehicles, based on the analysis of the experience gained in operation of manned submersible vehicles. The author of the review concludes that no general-purpose vehicle is feasible. Rather, each vehicle should be tailored to its specific task. The crew must be provided with maximum possible comfort, and all the information must be recorded automatically.

III. UNDERWATER SURVEYS

Underwater research work currently conducted throughout the world falls into the following few major categories.

Underwater expeditions with vehicles of various types.

The already-mentioned bathyscaph *Archimèdes* submerged in the Kurile Trench to a depth of 9545 m, near Puerto Rico to 8500 m, and in the Mediterranean to 5150 m. Since 1961, it had submerged 50 times to 7000 m, and 3 times to depths exceeding 9000 m. This vehicle has large operational capabilities [39].

The submarine *Aluminaut*, with the crew of 6, submerged to the depth of 1875 m in the Caribbean. The crew surveyed, through portholes, the sea floor and the bottom sediments, observed the fauna and flora existing at great depth, and determined the depth of sunlight penetration.

An interesting experiment on joint operation of the mesoscaph *Ben Franklin* and surface ships was performed in the Gulf Stream. In this test the mesoscaph drifted under water, while the surface vessel conducted synchronous observations on a traverse across the Gulf Stream. In addition to visual observations and still and motion picture photography through portholes and with externally mounted cameras, the crew of the mesoscaph traced the currents and measured the temperature, salinity, and other physical and acoustical properties of abyssal water in the Gulf Stream. They also measured sunlight penetration, water transparency, and sound scattering at depth, and conducted visual geological surveys at six diurnal stations 600 m below the surface, studying bottom relief and analyzing sections through the sediments to a depth of 100 m, while the mesoscaph was situated 10 m above the bottom. Other studies dealt with the patterns of attenuation and reflection of sound signals in bottom sediments, and fouling of the hull and meters by sea organisms. The *Ben Franklin*, with its crew of three American researchers and two Swedish navigators, traveled a distance of 2400 km deep in the Gulf Stream from Palm Beach, Florida, to the latitude of Boston. At times, its drift velocity reached 4.5 knots, which was double the expected speed [45]. During this drift, some of the collected material was lifted from the mesoscaph, in special containers, to the support vessel. Among the noteworthy findings was the detection of significant internal waves within the Stream induced by sharp uplifts of the sea bottom, as well as of giant submerged coral reefs, rising 30 m above the floor.

Soviet expeditions to Arctic seas, organized by the Polar Institute for Research in Fisheries and Oceanography (PINRO) and employing the *Sever-I* hydrosphere, accumulated much information on the behavior of fish in various subpolar fishing areas [9].

The results obtained so far in underwater expeditions indicate the need for long-term observations in deep waters and on the sea bottom.

Underwater stations and laboratories. The many available designs fall into two categories, namely mobile vehicles and stationary units.

The mobile vehicles include platforms of two types: (1) those intended for recording isotherms at depths to 1800 m; in this case the platform is submerged to the desired level and then drifts with the current at that level; (2) those intended for determining the distribution of water temperature to depths of 900 m; in this case the platform submerges automatically to the desired depth and then is made to rise to the surface. The projected work schedule for this platform envisages several 8- to 10-hour submergence–rise cycles of independent operation. All the movements of the platform would be controlled by a system of hydraulically operated servomechanisms coupled to electromechanical relay devices [87].

The stationary units are capable of much more diversified work, as seen from the examples cited below.

The underwater laboratory *Helgoland* was set up in the North Sea, 3 km to the east of the Helgoland Island at the depth of 22 m. It measured 9 m in length and 2.5 m in diameter, and had a self-contained air-supply system. It also carried a 14-day food supply. Unlike the 35 other underwater laboratories tested elsewhere in the world, this one was operated in the cold and rough waters of the North Sea. The first experiments involved the measurement of water temperature and current speed in the near-bottom layer, study of bottom sediments and plankton, and analysis of possibilities for breeding valuable fish, crabs, and molluscs in underwater oceanaria [80].

The West German biological underwater station *BAH-I* was tested in the western part of the Baltic Sea to determine the conditions that such a station would encounter during operations in the Baltic and North Seas.

The underwater habitat and laboratory *Bacchus*, capable of operating at depths to 180 m, can accommodate a crew of 6 for a week and conduct research at depths inaccessible to unprotected men [41].

The design of the Soviet underwater laboratory *Bentos-300*, used for specialized fisheries studies, incorporates the experience gained on previous underwater vehicles, in particular on the previously mentioned *Sever-I* hydrostat [9].

A vast and varied research program has been undertaken by the *Chernomor* underwater laboratory, operated by the Southern Branch of the Institute of Oceanology of the USSR Academy of Sciences [see N. A. Anbulatov, A. S. Monin, and V. S. Lebrebov in Okeanologiya **10**, No. 4 (1970); English transl. Oceanology **10**, No. 4 (1970)].

The Chicago Institute of Technology has carried out a design study for a system to be used in investigating the conditions for manned underwater work on fish breeding, oil and mineral mining, and development of deep-sea communications [21].

Planning has started on the design of a deep-sea base for the underwater mining of minerals and for carrying out other investigations at depths of as much as 500 m [22].

Aquanauts. The present and future use of deep-sea facilities calls for reliable life-support systems, and work is now being done on such systems as part of underwater expeditions and tests of stationary units. For instance, French oil and diving companies have run the *Janus* experiment which included a series of deep (down to 150 m) dives in a low-pressure chamber designed to determine the possibilities of work under water [49].

The *Textite-1* and *2* experiments run by the General Electric Company for the same purpose involved 2 months of work by aqualung-equipped divers at depths of the order of 15 m. This and many other experiments led to systems for testing man's capacity to work under deep-water conditions. The hazards due to deep-water submergence were determined, and the potential for communication between divers, of great importance in work in deep water, was studied as well.

An important aspect of study of the deep-sea regions is reconnaissance, and special facilities are being developed for that. For instance, deep-water observations are conducted by the Atlantic Institute for Fishing and Oceanography Research (AtlantNIRO) to check the work of trawl fishing gear with a towed deep-sea bathyplane. The data thus derived on the operation of the trawl include the shape of the net and its position relative to the sea bottom [11].

The Westinghouse Corporation has developed an unmanned, towed, submersible deep-water instrument carrier (in-air weight of 2950 kg) for studying the ocean floor. The carrier incorporates a side-looking underwater radar, movie and television cameras, and pulsed illuminators. It is intended for work at depths down to 6 km. The cigar-shaped instrument carrier is attached to a coaxial reinforced cable and is towed by a surface vessel at speeds of up to 1.5 knots. The movie and TV cameras are capable of scanning the sea floor in a band 30 m wide, with the carrier 10 m above the floor [67].

The vigorous and variegated underwater research now being conducted is given good coverage in the press, the greatest importance being attached to the projects concerned with the utilization of ocean resources.

IV. METHODS AND PROCEDURES

The techniques and procedures for collecting data in the ocean exhibit some specific characteristics. Thus there is a trend away from studies of steady-state conditions and toward the study of time-dependent process variations, a trend much influenced by new meteorological theories on the temporal fluctuations of atmospheric processes. However, most conventional oceanographic instruments are of little use for studying such time-dependent fluctuations in the flow of sea water. Hence, the development of instruments that will provide numerical information in Eulerian form is urgent. The instruments already available are current meters and temperature sensors attached to anchored buoys [73]. The buoy designs fall into three groups: stable buoys of the *Flip* type, low-draft buoys that fluctuate with the wind and waves, and intermediate buoys. Each group has its advantages but all are difficult to maintain in long-term operations [85].

Spar buoys with fiberglass floats, which carry a container with two inclinometers, a gyroscope, three accelerometers, and a compass, so that six degrees of freedom of the buoy movement can be measured, are available [83]. The multichannel recorder that complements these sensors is capable of operating for 40 min.

The equipment is switched by command from the service ship via a hydroacoustic channel. One such buoy was set in place in winds of 15 m/sec and waves 1.5 to 1.8 m high. A shipboard recording device is connected to the container by means of a zero-buoyancy cable. The upper part of the mast carries a bracket with an anemometer and a thermometer.

The *Triton* spar buoy is designed for studying air–sea interaction [84]. It consists of four pipe sections, a float, and a skeleton-type mast. Its total length is 34 m. The two lower pipes act as ballast tanks, the middle one holds the instruments and the power supply, and the upper is a float and a cylinder for the compressed air used for purging the tanks. Mounted on the mast, at heights of 6 and 10 m, are thermometers, a wind vane, a rain gauge, and a string wave meter; at the depth of 2 m there is a thermistor. The instrument container holds two inclinometers, an accelerometer, and a compass. The digital output of these instruments carries data on the buoy movements and is recorded by two tape recorders. The buoy is towed horizontally and righted by filling the tanks with water. The buoy was emplaced for 60 days at the depth of as much as 2.6 m. It operated in heavy seas (wave height to 4 m) and high winds (20 m/sec), the amplitudes of vertical and horizontal motion being 17 and 71 cm, respectively, and the heel angle reaching 2.5° [84].

The employment of buoy stations raises the problem of communicating with them over great distances. Very interesting experiments on long-distance buoy interrogation were done with buoy stations anchored in the North Pacific. In this case the *BRAVO (C)* station was used for daily transmission of data, via telemetry, to a shore-based transceiver located in California. The buoy station was equipped with a special computer that controlled the operation of the equipment, stored and transmitted the information via a radio channel. The telemetry system was shore-controlled (either automatically or manually). Interrogation and transmission of data occurred every 6 hours, and 174 communication experiments were carried out at optimal frequencies from August 1968 through March 1969. Of these transmissions 170 were error-free, but more errors occurred in transmissions at frequencies other than optimal [32].

A five-channel oceanographic telemetry system intended for data transmission from radiohydroacoustic buoys to a ship or shore information-collecting and -processing center was developed in Australia. Acoustic information is transmitted via four channels, the fifth channel being used for transmitting temperature and pressure data. The information can be transmitted over distances of 30 miles when a shore-based antenna is used, and over 5 miles when received by a ship. The buoy transmitter is capable of operating unattended for at least a month. A unit is provided for statistical workup of data [44].

Time-dependent fluctuations of oceanographic variables are best studied by a network of unmanned buoys rather than individual buoy stations. Such networks are capable of providing most of the desired information and do so cheaper than other systems. The National Data Buoy System now being planned in the U.S. is designed to provide careful analysis of the data from buoy-mounted sensors. The project involves development and pilot testing of a prototype buoy network and improvement of buoy techniques and sensors. About 100 American organizations participated in formulating a list of 69 variables, describing the state of the ocean and air, that shall be measured by the new network [85]. The project will be completed in several stages. The first stage will involve installation of a local network of 35 buoys for perfecting communications, methods of collecting and processing information, and checking out the buoys and their equipment. The global buoy network is to be subdivided into 7 oceanic and 6 nearshore regions, with a total of 540 stations, including 261 in deep-sea areas. The nearshore buoys will be spaced 180 to 300 km apart, and those in the ocean as much as 1100 km apart. The system will be deployed in 1975. The deployment of this system raises difficult problems concerning processing of the data (particularly of the vectorial values) for the purpose of extracting various frequencies from the full spectrum of temporal fluctuations, interpretation of the measurement, and of judgment as to whether they are representative of large areas of the ocean [86].

Even if the American system of telemetric buoys covering vast expanses of the ocean were available, the collection of masses of oceanographic data over the entire ocean can be accomplished only by joint efforts of several countries. With this objective in view, a working committee (USSR, Great Britain, France, West Germany, USA, and others) concerned with the establishment of an Integrated Global Ocean Station System (IGOSS) was convened in Paris in April 1968. The IGOSS will provide regular information on the condition in the hydro- and atmo-sphere. It will comprise national facilities and services whose activities will be coordinated by UNESCO, World Meteorological Organization (WMO), and other international bodies. The IGOSS should work only for peaceful purposes. The system is to comprise a network of stations, including automatic telemetric buoys, coastal stations, research vessels, fixed ocean stations, nearshore stations, and data-collecting centers [59]. The synchronized information on the state of the sea and air above it will be used for current decisions and for research.

Another characteristic feature of modern oceanographic research is expeditions to special test areas which are outlined either in the open ocean, or in the nearshore zone, where a coastal base is created. In the former case a ship performs measurements on a contour enclosing a relatively small representative area of the ocean. Such work is performed by ships of many countries, including those of the USSR. The US Navy employs several nearshore sea test ranges. A special oceanographic test range for commercial organizations is to go into operation in Hawaii. It will have 18 oceanariums, laboratories for underwater physiology, bioacoustics, and sea-pollution fisheries-damage assessment, as well as repair shops and offices. It will have a low-pressure submersible chamber for pressures of up to 60 kg/cm^2. A 70 mile-long ocean route for research and testing of equipment has already been equipped. The test range facilities will include 4 ships, 2 deep-water vehicles, and an underwater house for a crew of 6 to work with drilling units at a depth of up to 200 m. The test range will train personnel and will attempt to find ways to shorten the time needed to design equipment, particularly that intended for the deep-water drilling for oil and gas [66].

Finally, wide use is made of essentially new means for conducting oceanographic observation. Thus, aircraft are being used for oceanographic survey, delivery, setting up and recovery of various instruments, and for collection of samples. An American company is developing equipment for infrared sea-surface temperature measurements by taking samples from a slow-flying aircraft (one such system has already been tested). The equipment includes an aircraft winch, a cable, a drop probe with detachable ballast and a recorder. A system is being developed for direct measurement of water temperature from aircraft flying at speeds of up to 330 km/hour at altitudes of up to 300 m. During measurements, the cable is unrolled from the winch and is trailed as a loop in the air behind the aircraft. Then the probe is dropped. As soon as it hits the water, the ballast is detached. The probe thus stays on the surface, to be retrieved 5 to 6 sec later by means of the cable that is winched aboard the aircraft. The cable is 7.6 mm in diameter; its tensile strength is 225 kg with 100% elongation. It has a nylon core, with a conductor wound spirally round it. The probe is an aluminum cylinder with an inside thermistor gauge. It is believed that it may be possible to take water samples with such a probe. The airborne recorder associated with the probe is based on a Wheatstone bridge principle [30].

A biotelemetry system employing dolphins as instrument carriers is of interest. A radiotransmitter having a neutral buoyancy is attached to the dolphin's dorsal fin. The transmitter is activated only when the dolphin surfaces to breathe. The dolphin's travels were thus traced for 2 weeks near California. Transmitters are now being developed for telemetry of water temperature and the depth of submergence of the dolphins. Biotelemetry will be used to study dolphin physiology under natural conditions, as well as physical parameters of the environment [27].

Radar is used as a research tool in oceanography. In determining the sea state, the best results are obtained with pulse durations of 10 to 20 graduations. When the radar antenna is set up low over the sea (so that only wave tops can be observed), medium-length waves can be measured under the conditions of heavy seas. Other parameters can be obtained from Pirson's formulas. Doppler radar with both vertical and horizontal polarization allows determination of the effect of the wind on the surface water (translational motion), as well as wave heights ranging from 1 to 1.2 m. Radar is expected to become a technique for studying the development and decay of waves [65].

Soviet researchers from the Marine Hydrophysical Institute of the Ukrainian Academy of Sciences have developed a controlled system of towed hydrophysical instrumentation for measuring temperature distribution (in the range of 2° to 35°), salinity (electroconductivity 17 to 71 mS), hydrostatic pressure (0 to 40 atm), and vertical temperature gradient (10°/m) to a depth of 300 m, the towing speed not exceeding 12 knots [17].

A quartz thermometer is used for precision measurements of water temperature at great depths. The system comprises two sensitive temperature probes, a set of transformers, and recorders with a digital counter [60].

The *Molodezhnaya* Arctic station uses an all-weather electronic tide-gauge attachment, linked to a recording potentiometer through a 600 m-long four-core cable [7].

The above methods and instruments mark a new stage in sea-observation technology.

V. SATELLITES

Space technology is important to oceanography as it offers the capability of global study of various processes in the ocean and of long-term forecasting.

Earth satellites are used to study the sea state, tsunamis and long-term fluctuations of the ocean level, currents and surface-temperature distribution, as well as for detection of accumulations of plankton, fish, and other sea organisms, bioluminescence, fresh water, drifts, wastes, icebergs, location of storms, fogs, reefs, sunken vessels, etc. Satellites can be equipped with microwave radiometers, scatter meters, and photometers for the visible, infrared, and microwave ranges. Such instruments (plus various techniques such as movie films, radar, spectroradiometry, and polarimetry), carried by polar- and equatorial-orbit satellites ensure all-weather, round-the-clock collection of meteorological and oceanographic information over the entire ocean. Apart from the need for new, satellite-borne equipment, there is a need in satellite-based oceanography for wider application of contact measurements by means of buoy stations and surface ships. This will yield reference data for use in interpretation of satellite observations. There is also a need for new systems to process the vast amount of satellite-derived analog, digital, or graphic information.

Using radar, laser, infrared, film, photographic, and television instruments carried on satellites of the *Nimbus-2* series, the U.S. Naval Oceanographic Office has conducted experiments on measuring the surface ocean temperature, sea state, surface-wind velocity, detection of submerged reefs, shoals, and fish schools, and observation of ice state. The infrared equipment was used to study the temperature of the Gulf Stream. The Gulf Stream temperature gradients, computed from data obtained by satellites, ships, and aircraft, were in good agreement for the position of the thermal axis of the Gulf Stream axis, but the isotherms of the marginal zones of the current, as measured from satellites and from ships, disagree. Satellite-derived temperature data must be smoothed out before such comparisons can be made, since they are distorted by the cloud cover and the inherent noise of the system.

The sea state was observed with radar suitable for measurement of waves of 1.5 to 8 m in height. Laser-equipped instruments allowed observations of the state of the sea from altitudes of 2100 m in daytime and 4500 m at night. Depth could be measured to 50 m with the accuracy of ±0.45 m. This equipment was used, among others, to find submerged reefs in Florida, and may find application in detecting fish schools, determining variations of temperature gradients with depth, etc. The use of satellites for ice reconnaissance in the Arctic and other regions is being discussed. Special attention is given to developing the techniques and equipment for all-weather measurements. For example, an all-weather radiometer has been developed [62]. The US conducted a full-scale simulation test of a system intended for collecting geophysical data by means of buoy stations and relaying them with the help of the *Nimbus-B* weather satellite. The tests led to the development of a simulation device for spotting buoy stations in the ocean and determining their locations to within ±1 mile, interrogating the stations and recording the data [26].

Soviet researchers concentrate on methods of processing satellite photographs and utilization of the satellite data for determining ice conditions in the Arctic (for example, the boundary lines between ice areas of various densities). There is a fairly good agreement between 10-day ice-forecasting maps derived from satellite and aircraft photographs [18].

The employment of satellites for studying the ocean has led to the development of a new branch in the science of the sea, namely satellite oceanography. Further progress in this area depends largely on improvement of observational equipment. Full utilization of the potential of these new tools will require coordination of international scientific studies.

VI. DATA PROCESSING

A vast increase in the data has increased the attention to mechanical processing of oceanographic information. Thus a system for observation processing using the M-20 computer has been developed at the Department of Oceanology of the Leningrad State University. The system rearranges the data into a form convenient for further processing, samples them in terms of given features and stores them on magnetic tape. A computer algorithm for selecting series of hydrologic characteristics has been developed.

A system for automatic processing of oceanographic data has been developed in Great Britain and installed on some research vessels. The system receives the input data from telemetric buoys, and does the primary computer processing, with the output (on punched tape) of the corrected data being transmitted for direct use at sea or further processing on land.

Canadian technicians have developed the *Nuro* system for rapid data processing, which can be installed either on board a ship or on land. Information can be collected by any vessel and data are fed into the system, whose output consists of graphs and tables showing the position and course of the vessel and corrected sounding data.

VII. NEW OBSERVATIONAL EQUIPMENT

The emphasis in ocean research is on automatic, self-contained electronic and telemetric devices, towed or lowered with the help of cables, and intended for continuous recording of oceanographic variables.

An automatic drop sonde for measurement of vertical temperature and salinity distribution down to depths of 1000 m during free drop and surfacing was developed in the U.S. The 1.5 meter-long sonde submerges under the weight of a load, which is detached automatically at a preset depth, after which the sonde surfaces. A digital tape recorder is used to record the data. Electroconductivity is measured by means of a variable inductor, temperature is determined with a thermistor, and the depth by a spring-actuated pressure meter, each variable being recorded every 2 min in a form suitable for direct computer input [74].

Deep-water devices for measuring temperature, salinity, and depth were developed by Bissett–Berman Corp. These devices are a salinometer, a detachable bathythermograph, and a multiple sampler. The sensor in the automatic temperature-recording system is a platinum resistance thermometer. The recording salinometer determines salinity *in situ*, a built-in analog computer providing automatic temperature and pressure corrections for the salinities measured in terms of electroconductivities. These instruments can operate to depths of 6000 m. Each instrument carries a float and a weight, the latter being detached when the device touches the sea floor. The instrument is lowered by simply throwing it overboard [78].

A novel system for recording water temperature, speed and direction of current is capable of operating at depths of 300 m. It employs a thermistor (range 0 to 40°C), a rotating-vane meter for measuring current speed (0- to 0.5- and 0- to 3.5 m ranges), and a vane connected via a clutch to a tracking potentiometer and to a compass for determining current direction (0 to 360°). The data are fed to a single-channel recorder [47].

There exist similar devices for measuring pressure, temperature, and currents in the near-bottom layer at great depths. The pressure meter in this case is of a reed type and can operate at a depth of up to 5500 m, with a resolution of ±0.5 mm. The temperature sensor comprises two quartz oscillators, a mixer, and an output amplifier, the sensor resolution being at least 10 ppm. The current speed and direction in the near-bottom layer is measured with three pairs of thermistors, the signals from these being fed into a computing device whose output is a direct current velocity and direction reading. The data are recorded in digital form. The depth, temperature, and current meters, plus the power supplies, are mounted on a special frame placed on the bottom. Two aluminum spheres of positive buoyancy are connected to the frame by a cable. They house digital recorders, acoustic devices, a radiobeacon, and devices for light signaling. The unit is lowered by a ship to the sea floor where it remains for several days or weeks. At the end of the period, the ship returns to the place where the unit was dropped, an acoustic signal from the ship triggers a release mechanism, the submerged unit drops ballast and surfaces. The unit was lowered to the ocean floor about 40 times and yielded valuable information on thermal and hydrodynamic properties of the near-bottom layer and on the tides in the open ocean [28].

An automatic deep-water unit for oceanographic research, comprising a set of meters and sensors, was developed at the Scripps Institution and has produced good results. The unit incorporates a trap for sea organisms (capable of operating to depths of 5500 m), a device for taking samples near the bottom, a recording current meter (to 8000 m, range 0.3 to 50 cm/sec, direction-measuring accuracy of ±5°, length of self-sustained operation 20 days), and a still camera with a flash. It is housed in a radiobuoy with a signal device, gasoline-filled tanks, meters and sensors, a disconnecting device, and an anchor [75].

A controllable towed body for oceanographic research, the *Delphin*, was developed by the German Hydrographic Institute. When ship-towed, the device follows a sawtooth path in the vertical plane, obeying signals fed by cable, or it can move horizontally at a preset depth under manual or automatic control. It can carry instruments for measuring the following combinations of variables: pressure, seawater conductivity, and temperature; or pressure, temperature and sound velocity; or pressure, temperature, and turbulence. Its operational depth range [31] is 0 to 100 m.

An American company has developed a telecontrolled, high-speed oceanographic research surface vehicle. The vehicle is 4.4 m long, its draft is 0.7 m, its speed is 3 to 20 knots, and it is capable of operating at speeds of 15 knots for 20 hours. The vehicle is provided with a sounder, a towed thermometer, hydroacoustical and telemetry systems, and other meters and sensors.

The Marine Hydrophysical Institute of the Ukrainian Academy of Sciences has developed a unit for measuring the main hydrological variables, comprising a buoy, a towed system, and a probe system. Its central unit is an analog-to-digital converter with digit-by-digit balancing. The probe system has three measuring channels, the towed device has four channels, while the automatic buoy station has six. A platinum thermometer is the temperature sensor, the depth sensor is a spiral tubular spring whose pressure-actuated motion varies the resistance of a potentiometer. The electroconductivity is measured by a variable inductor, which is essentially a transformer bridge. The towed device is provided with thermistors for measuring the temperature gradient. Current direction is determined by a vane and a compass [8].

The increasing use of towed devices has produced a need for instruments capable of accurate determination of the depth of submergence of the tow device, depending on the tow speed, the cable diameter, the drag, and other factors. A variant of a towed instrument, namely a self-contained buoy for simultaneous measurements of water temperature at four levels, was developed at the Applied Physics Laboratory of the University of Washington. The device comprises a cable-supported instrument capsule and four external ±0.0005° quartz temperature sensors spaced vertically at 5 m intervals. The capsule holds measuring and control instruments, a seven-channel digital tape recorder, and a power supply that allows the system to function independently for 24 hours [68].

The Bedford Institute (Canada) has announced an automatic bathythermograph for operation at depths to 210 m while stationary and in the range of 18 to 60 m when towed, the data being transmitted to the towing ship by cable [58].

An American *in situ* salinometer for use in the deep ocean employs two coaxial toroidal transformers as salinity sensors. The thermal compensation system employs two platinum thermistors which form a double bridge calibrated for 35‰ salinity. The system output depends on water salinity, and phase-frequency modulation is employed in data transmission. The rated salinity-measuring error is ±0.02‰ if the salinity is in the range of 30 to 40‰, temperature is between 0 and 30°, and the pressure is between 1 to 700 kg/cm^2. About 90 such instruments are now in use and show good accuracy and reliability [23].

The conventional hydrological devices employing sampling bottles have been improved at the Lamont Geological Observatory. The new device is essentially a sampler lowered together with a probe on a cable that carries the electrical signals which trigger the mechanism reversing the thermometers at the required depth while closing the bottle. This automatically gives the temperature corrections for the probe readings.

Highly stratified waters can be sampled with a horizontal sampler, which is a 1.6 liter PVC cylinder open on both ends and fitted with hemispherical lids. The lids are held in the open position by chains whose ends are attached to a trigger mechanism. The sampler can be used in water layers [38] as thin as 0.1 m.

A new marine geological device photographs a 60×60 cm area of the sea floor at a depth of 6000 m and takes a ground sample in the area photographed. This is essentially a grab fitted with a camera and a flash at its sides. They are triggered at a preset distance from the bottom [20]. There also exists a remotely controlled submerged instrument carrier working in conjunction with a ship and performing automatic sounding of the sea floor on a track parallel to the ship's course and several miles away from it. A single ship can tend several such carriers [58]. A French-made unit for automatic survey of bottom topography variations due to the erosion of bottom sediments allows forecasting the possible zones of erosion and deposition, by modeling the results obtained on actual shoals in a 60×30 m test basin [51].

To examine the phenomena occurring near the bottom, a special carrier has been equipped with instruments to measure the average current speed, two still cameras, a special cone-shaped thermometer–receiver capable of sampling the bottom sediments and carrying a device for sampling water with bottles at different depths [69].

Among other promising automatic devices are those which can be towed and submerged to preset depths to take a set of measurements, with automatic recording.

VIII. SEA EXPEDITIONS

Sea expeditions are the main source of information about the ocean. Such expeditions are mounted by many countries, often fit approved international programs, and are sent to various areas of the ocean.

For example, Canada is engaged in its first global oceanographic venture, the Hudson-70 project, aimed at exploring large areas of the ocean by means of the R/V *Hudson* (displacement 4800 tons, speed over 16 knots; the hull is reinforced for travel in ice; the deck accommodates a helicopter and several launches for shallow-water operations). The 41,000-mile cruise is designed to traverse the Atlantic and Pacific Oceans and the seas near the Canadian north coast. The program envisages research in biology, chemical and physical oceanography, geophysics, hydrography, and underwater acoustics [43].

Interdisciplinary sea expeditions normally explore comparatively small areas of the ocean. Thus, between February and May 1969 the Soviet R/V *Akademik Kurchatov* explored areas in the west-central Atlantic Ocean, investigating the current dynamics at the margins of the western tropical zones, variations of the hydrochemical and hydrological fields, physical processes in the atmospheric boundary layer, the geophysical fields of the Earth, sea-floor topography, as well as geological and biological processes. The most important result of the mission was the discovery of a long (over 3500 miles), strong (1 to 1.5 knots) current, countercurrent to the Antilles and Guiana Currents. The crew of the *Akademik Kurchatov* also tested new instruments for chemical and gravimetric investigations, as well as for cosmic ray observations. The vessel carried a Minsk-2 computer and the cruise proved the value of an on-board statistical and computation team [12].

The equatorial zone of the Atlantic was investigated by the *Akademik Vernadskiy*, another Soviet research vessel [3].

Three successive oceanographic surveys under the *Northwestland* program were conducted in the northwestern Atlantic. The observational data obtained by the vessels engaged in the survey were supplemented with information derived from weather ships and other vessels operating in the Irminger Sea, in the Labrador Basin, and the Strait of Davies. Three new types of water were discovered. The results of the survey were used to compute the variations in the conditions that occurred during the period of the survey [52].

Expeditions of the Institute of Biology of the Southern Seas to the Adriatic, Ionian, Tyrrhenian, and Ligurian Seas and the Strait of Tunisia studied the biology of fouling, as well as the radiobiology, hydrology, and hydrochemistry of these waters. Measurements were taken at 53 sea stations.

A special expedition of the R/V *Discoverer*, from January through March 1968 was designed to find out whether the contours of the continental shelves and the continental slopes of South America and Africa are in mutual conformance down to the 1800 m isobath in area between Dakar and Abidjan. The expedition program included sounding, meteorological, aerological, hydrological, and geographical observations, dragging, sampling, and photography of the sea floor. To establish the geophysical similarity of the two continents, seismic, magnetic, and gravity surveys were made. The Kayar sea canyon (on the coast of Senegal) and a canyon in the Abidjan region were studied in detail.

In recent years the Australian R/V *Diamantina* has been surveying the Indian Ocean west of Australia, mapping its hydrology and hydrochemistry and performing other hydrobiological and geological work [55, 55a, 55b, 55c].

Of the many expeditions in the Pacific, that of the R/V *Coriolis* is of special interest. It carried out observations on an equatorial traverse between the Galapagos and the Solomon Islands. The data obtained allowed a more accurate mapping of the vertical boundaries of the Cromwell Current. The upper boundary of this countercurrent is clearly revealed over its entire length by the high temperature, density, O_2- and phosphate-content gradients. The depth of that boundary, which is 130 m in the western part of the ocean, gradually decreases to 30 m at Galapagos Islands. The lower boundary shows up more clearly in the data from the western than from the eastern part of the ocean. The mean thickness of the Cromwell Current is 250 m.

Important hydrological results were obtained by a West German expedition to the Kuroshio Current area. It was found that the temperature of the surface water in the Kuroshio Current always exceeds 15°, the prevailing temperature being 20° or even higher. Generally, the same temperature is observed to depths of 100 m and more. The Kuroshio Current is wavelike, with the wavelength of 80 to 240 nautical miles. In the area where Kuroshio branches off into Kumanonada Bay, the 15° and 18° isotherms converge under the influence of the surrounding cold water [40].

Some interesting methods and procedures were tested by the Japanese ship *Ryofu-maru* on cruises to the west of Kyushiu Island on three traverses across the warm Tsushima Straits.

Investigation of the Arctic and Antarctic Oceans were continued. For example, in early 1969, the area to the north of Alaska was investigated by the American icebreaker *Staten Island*. The prime objective of this mission was to determine the strength of the pack ice to see whether merchant vessels could pass through it. It was found that the opening of the north coast sea route for year-round use would require new, more powerful icebreakers.

Oceanographic conditions of the western part of the Pacific sector of the Antarctic were investigated by the R/V *Umitaka-maru* of Tokyo University. Observations were made at 70 successive stations, and 1 to 2 bathythermograms were taken in the intervals between the stations for a total of 197 bathythermograms. Generally, the conditions were investigated at 22 standard levels. The conditions in the zone of Antarctic convergence were analyzed using data on the distribution of the subsurface temperature and salinity to the south and north of this zone. The data on water temperature, salinity, and O_2 concentration were used to define the flow of Antarctic water (winter flow, intermediate and near the bottom) and abyssal water, which is warmer. Temperature variations in the 0- to 20 m layer are due to time variations of insolation and air temperature, but these factors never affect the Antarctic winter flow. Salinity of surface waters was somewhat lower in the area of the southernmost stations (70° S). The temperature and salinity of the Antarctic winter flow is fairly stable, particularly so in its core and the upper levels.

In the layer intermediate between the bottom of the Antarctic winter flow and the top of the abyssal warmer water, the temperature and salinity vary considerably. The heat exchange between these waters was determined. It appears to be affected not only by the rate of rise of the warmer abyssal water, but also by the lateral advection [46].

Joint international oceanographic expeditions work in Antarctic waters, just as in other sea areas. Thus, in early 1969, the Weddell Sea was explored by the American vessel *Glacier*, in cooperation with the Argentine icebreaker *General San-Martin*. Aside from their normal research program, the vessels continuously recorded the water salinity, temperature, and depth. The preliminary results are as follows. There are large variations in temperature and salinity between depths of 0 to 50 m, due to local ice conditions. A layer of cold winter water was detected between depths of 50 to 75 m, while warmer water exists between 250 and 500 m. The bottom Antarctic water (salinity of 0.6 and 34.66‰) was found at depths of over 3500 m. The warm water found in the 100-mile zone to the north and northwest of Cape Norway is thought to be an intrusion of the West Wind Drift [29, 29a].

An international expedition exploring the Kuroshio and adjacent areas began its work more than five years ago. The 1965-1967 research resulted in detecting new features of water circulation in the area, in defining its seasonal and annual variations, and establishing the role of eddies in generating water circulation in the front zone and the inner area of the subtropic anticyclonic gyre [14].

From January to March 1968, observations were made on four traverses east of the Philippines, along the 127°, 130°, 133°, and 137° E meridians. The variations in hydrological conditions in the western Pacific during the 1967 and 1968 winters were charted. The Equatorial Countercurrent was mapped at the latitudes of 1° to 2° N 200 to 300 m below the surface. Its velocity was 1 knot in 1968 and ~3 knots in 1967. The South Equatorial Current (velocity of ~2 knots), carrying salty tropical waters from the southern Pacific, was observed near the Equator at depths of 100 to 300 m. The flow of the 35.5% salinity waters was larger in 1968 than in 1967. To the southeast of Mindanao, the currents form a cyclonic eddy, [54] in which the maximum speed is 3 knots at a depth of 600 m.

The International Indian Ocean expedition (IIOE) has generated information for important scientific generalizations. Thus, the 18,000 water analyses made yielded sufficient data for outlining the hydrochemistry of the Red Sea and the Gulf of Aden. It was found that only 30% of the south-to-north increase in the salinity of the surface water of the Red Sea was due to evaporation, with the other 70% due to inflow of abyssal water from depths of the order of 1500 m. The data on the distribution of the dissolved oxygen along the Red Sea axis were used to model the meridional circulation in that sea and to explain the distribution of phosphates, nitrates, and silicates in it. Data on the "lifetime" of abyssal water were used to calculate the silicate concentration norm (1 ng-atom/liter). It appears possible to calculate the "lifetime" of water outside the Red Sea from its silicate concentration. Higher silicate concentrations and higher oxygen deficiency are concomitants of longer "lifetimes" below the thermocline (48 years instead of 30).

The data on the vertical distribution of the dissolved inorganic phosphates were used to calculate the vertical exchange (1 to

4 g/cm · sec) and the velocity of vertical flow (10^{-5} to 10^{-4} cm/sec). The distribution of phosphates, silicates, nitrates, and NH_3 in the Red Sea and Bab el Mandeb waters was analyzed, and the silicate–salinity ratio was suggested as a characteristic of the Red Sea water. Expressions for calculating the various components of the carbonate system, ion activity, and $CaCO_3$ content in these waters were derived, allowing for the effect of temperature and pressure. From the data on $CaCO_3$ sedimentation [36] the overall rate of sediment accretion in the Red Sea was computed as $2 \cdot 10^{-3}$ cm/yr.

International oceanographic expeditions have been working in the Baltic Sea since early 1969, in accordance with a plan of the International Council on Baltic Sea Exploration, adopted in Copenhagen in the fall of 1968. Six countries, including the Soviet Union, are to carry out (in succession) 3-week measurements of salinity, water temperature, and other hydrochemical characteristics in the Baltic Sea (except the Gulf of Bothnia). The information thereby gained will be used for improvement of forecasting methods [57].

Thus oceanographic expeditions now engaged in integrated and specialized exploration of the ocean are not only scientifically but also practically oriented.

IX. AUTOMATION OF OBSERVATIONS AND DATA PROCESSING

The introduction of modern electronics and communications into oceanographic practice has ushered in a period of fundamentally new methods of collecting and processing observational data, employing large instrumental and computing systems.

Design bases for large data-collecting and -processing systems for oceanographic research, capable of collecting very large amounts of information and of multidimensional control, measurement, and analytical operations are now being developed.

Some of these developments have now been incorporated into practical system. Thus the Marine Hydrophysical Institute of the Ukrainian Academy of Sciences has produced an automatic system comprising a measuring unit, a data-acquisition and -transmission module, and an on-board computer center. One such computer center, on the R/V *Mikhail Lomonosov*, is built around the Dnepr-1 computer, and is capable of performing statistical analysis of scalar and paired observation series, analysis of current data, reception and deciphering of hydrophysical data, correction of the ship's course, determination of the ship's most likely coordinates, and analysis of depth measurements. The computer printout is in a form suitable for automatic document storage and retrieval. The same Institute has developed a data-handling system employing punched cards operating in conjunction with visual checking of data stored in the magnetic disk operational memory of the Dnepr-1 computer.

The shipboard computer achieves initial analysis of oceanographic information. Further processing of the data is accomplished in a shore-based computer center where the entire available information is used to plot the analytically derived fields of hydrophysical characteristics [2].

Some experience on automating oceanographic research has been gained by the crew of the R/V *Akademik Kurchatov*. During its third voyage, they employed an electrical thermograph for studying temperature distribution in the water layer extending from the surface to 200 m. The temperature and salinity data were fed to the Minsk-2 computer which can fully analyze a series of 1000 vector terms in 40 min. On-board equipment allows direct input of analog signals to the Minsk-2 digital computer via a 24-channel analog-to-digital converter; alternately, the analog information can be stored in a buffer memory. The digital information can be fed to an X–Y graph plotter, which can plot a graph in 20 min. The on-board computer is also used for navigational and hydrographic calculations and for generating calibration tables for the solid-state temperature sensors [16].

In addition to on-board units, the Marine Hydrophysical Institute is working on controlled automatic systems for buoy stations. A command radio link for telemetric control of automatic buoy stations was developed. The system interrogates the desired buoy, activates its power supply, controls data collection and transmission, checks and calibrates the telemetric control equipment, and remotely tunes the buoy transmitter [4].

Self-contained experimental automatic measuring systems equipped with telemetry facilities have been installed on US Navy's floating buoys. The best results have been obtained with the NOMAD systems, which are boats ~6 m long and 3 m wide, displacing as much as 9 tons. The hull is aluminum and the power is generated by radioisotopes. The system needs servicing only once a year, and can operate unattended the rest of the time [56].

Much valuable data can be collected by measuring small-scale inhomogeneities at sea, using floating instruments whose parameters are independent of the time and position in space. Such instruments include the integrating photothermograph whose built-in strain gauge is connected in a bridge circuit, with the measurements being integrated over a time that is small compared to the time constant of the strain gauge. Hydroacoustical telemetry was successfully used in the Benthos model 1024 (USA). This self-monitoring instrument can record and transmit data on water temperature and its own depth of immersion, while a new method for computing salinity from specific electrical conductivity, temperature, and pressure was developed with the aid of the general-purpose Dnepr-1 digital copmputer. The experience gained was used as a basis of the recommendation of a design of a special-purpose digital computer [13].

To sum up the achievements in automation of hydrophysical research, let us point out that as many as 20 measuring systems have been already built for transmission and processing of information on the physical fields existing in the ocean. The descriptions of some of these systems were published [10]. Automatic systems for collection and processing of hydrophysical information are still in their infancy, but this is a very promising trend in oceanographic research.

LITERATURE CITED

The number in parentheses refers to the number of the citation in the Soviet Abstracts Journal (Referativnyy Zhurnal) Geophysics 1970.

1. Agenorov, V.K., Distribution of the deep-water hydrological stations in the ocean and the amount of observational data available on each area of the ocean. Trudy Gos. okeanograf. instit., 1970, No. 100, pp. 71-87 (1970, 10B11).
2. Belyaev, V.I., Computers and research on physical fields in the ocean. Visnik Akad. Nauk Ukr. RSR, 1969, No. 4, pp. 35-42 (1970, 4B16).
3. Boguslavskiy, S.G., A.Ye. Babinets, and A.P. Metal'nikov, Investigation of the near-Equator area of the Atlantic Ocean during the first voyage of the R/V *Akademik Vernadskiy.* In: Morsk. Gidrof. Issled., 1969, No. 2 (44). pp. 3-18, Sebastopol' (1970, 4B158).
4. Govalov, K.L., S.V. Dozenko, V.I. Katolichenko, and S.V. Kuleshov, Radiocontrol of automatic buoy stations. In: Avtomatiz. nauchn. issled. morey i okeanov (Automation of Research on the Sea), Sebastopol, 1969, pp. 49-54 (1970, 4B24).
5. Dolgopolskaya, M.A., and D.M. Vityuk, Investigations by the Institute of Biology of the Southern Seas in the Mediterranean. In: Ekspedits. issled. v. Sredizemnom more v sent.-dek. 1967. (Expeditionary Work in the Mediterranean in Sept.-Dec. 1967). Kiev, Naukova Dumka Press, 1969, pp. 3-18, (1970 4B158).
6. Dozenko, S.V., Measurements of hydrophysical fields with mobile instruments. In: Avtomatiz. nauchn. issled. morye i

okeanov (Automation of Research on the Sea). Sebastopol, 1969, pp. 87-89 (1970, 4B25).

7. Yeliseyev, V.V., Use of electronic attachment to a tide gauge. Inform. byull. Sov. Antarktich. ekspeditsii, 75, pp. 67-70, 1969 (1970, 5B41).
8. Zaburdayev, V.I., L.M. Antonova, N.N. Anblagova, L.M. Dobruskina, V.N. Meyerovich, and G.V. Smirnov, Instruments for buoy, towed and probe-type hydrological systems. In: Avtomatiz. nauchn. issled. morye i okeanov (Automation of Research on the Sea), Sebastopol, 1969, pp. 106-112 (1970, 4B26).
9. Kiselev, O.N., Construction of submersible laboratory capable of submergence to 300 m. In: Morskiye podvodnyye issled. (Underwater Research at Sea), Nauka Press, Moscow, 1969, pp. 301-302 (1970, 5B24).
10. Kolesnikov, A.G., and V.I. Belyaev, Some results of automation of hydrophysical research. In: Avtomatiz. nauchn. issled. morye i okeanov (Automation of Research on the Sea), Sebastopol, 1969, pp. 5-10 (1970, 4B19).
11. Korotkov, V.K., Bathysphere and its use for the observation of the operation of a bottom trawl. In: Morskiye podvodnyye issled. (Underwater Research at Sea), Nauka Press, Moscow, 1969, pp. 12-19 (1970, 5B26).
12. Kort, V.G., Main scientific results of the expedition of the R/V *Akademik Kurchatov* (5th voyage), Okeanologiya, 9, No. 5, pp. 910-916, 1969 (1970, 5B133).
13. Lyashenko, Ye.F., and M.G. Poplavskaya, Instruments for automatic calculation of the sea water salinity. In: Morskiye geofiz. issled. (Marine Geophysical Research), 1969, No. 2 (44), pp. 173-182, Sebastopol (1970, 5B56).
14. Makerov, Yu.V., Water circulation in the Kuroshio area from international research data, 1965-1967. Trudy Gosud. okeanograf. instit., 1970, No. 100, pp. 22-28 (1970, 10B162).
15. Marti, Yu.Yu., Scientific and fishing vessel *Akademik Knipovich*, Trudy Vsesoyuzn. n. i. instit. morsk. i rybnogo khoz i okeanograf., 66, pp. 13-20, 1969 (1969, 12B8).
16. Paka, V.T., L.M. Krivelevich, and O.A. Gushchin, Experience in automation of oceanographic research gained during the third voyage of the R/V *Akademik Kurchatov*. In: Avtomatiz. nauchn. issled. morye i okeanov, 1969, pp. 20-26, (1970, 4B21).
17. Paramonov, A.N., V.M. Kushnir, A.V. Khokhlov, and V.F. Zhuravlev, A system of controlled towed hydrophysical equipment. Trudy morsk. gidrofiz. instit. Akad. Nauk Ukr. SSR, 41, pp. 164-180, 1969 (1970, 8B24).

18. Sinyurin, Yu.N., Application of the information on ice obtained by satellites, Meteorologiya i gidrologiya, 1969, No. 8, pp. 94-96 (1970, 1B33).
19. Abel, R.B., Ocean science and technology. A review of the past; a glimpse of the future. NEREM Rec., 1967, 9, pp. 166-167, Boston, 1967 (1970, 3B8).
20. Alix, Y., A. Le Grand, and V. Romanovsky, La benne s caméra. Trav. Centre rech. et études océanogr., 8, No. 3, pp. 31-38 (1970, 4B57).
21. Benfield, B., and J. Kaczkos, Hydrospace. Marine Technol. Soc. J., 3, No. 4, pp. 45-55, 1969, (1970, 4B206).
22. Bruneau, A.A., Proposal for a Canadian underocean lab. Sea Harvest and Ocean Sci., 1970, Apr.–May, pp. 30-31 (1970, 11B17).
23. Brown, N.L., An *in situ* salinometer for use in the deep ocean. Marine Sci. Instrum., New York, 4, pp. 563-577, 1968, (1970, 6B82).
24. Busby, B.F., *Ben Franklin's* oceanographic mission. Undersea Technol., **10**, No. 5, pp. 64-67, 70, 72, 1969 (1970, 2B21).
25. Boylan, L., Soviet-block oceanographic instrumentation. Libr. Congr. Aerospace Technol. Div., Washington, D.C., 1969, 28 pp. Ref. U.S. Govt. Res. and Develop. Repts., 69, No. 14, p. 72, 1969 (1970, 6B12).
26. Bronzino, J.D., R.W. Corell, S.L. Heller, F.R. Hess, D.J. Meeker, and D.W. Melvin, Full-scale simulation of the IRLS buoy program. NEREM Rec., 1967, 9, pp. 194-195, Boston, 1967 (1970, 2B27).
27. __________. Bio-telemetry—using marine mammals to probe the sea. Trans. Inst. Marine Engrs., **81**, No. 6, 1969. Marine Engrs. J. June, pp. 10-11 (1970, IB36).
28. Caldwell, D.R., F.E. Snodgrass, and M.H. Wimbush, Sensors in the deep sea. Phys. Today, **22**, No. 7, pp. 34-42, 1969 (1970, 1B18).
29. Capurro, L.R.A., Research program in cooperation with Argentina. 1968-1969. Antarct. J. U.S., 4, No. 4, 98, 1969 (1970, 4B171).
29a. __________. International Weddell Sea Oceanographic Expedition. *Ibid.*, 4, No. 4, p. 96, 1964 (1970, 3B139).
30. Converse, C.B., and E.C. Brainard, II. Sea surface temperature verification from photometry data and surface water sampling from fixed-wing aircraft. Marine Sci. Instrum., New York, 4, pp. 296-300, 1968 (1970, 6B61).
31. __________. *Delphin,* a controllable towed body for oceanographic research. Donier-Post, 1969, No. 1-2, pp. 41-43 (1970, 5B36).

32. Devereux, R. F., Long-distance interrogation of ocean data stations moored in the North Pacific Ocean. Telecommun. J., **37**, No. 4, pp. 173-185, 1970 (1970, 11B114).
33. Felts, H. C., U.S. Naval Oceanographic Office—a review. Newsletter, Nat. Oceanogr. Data Center, No. 7-8, pp. 1-7 1969 (1970, 3B9).
34. Fujinami, N., Research fleet of the world. Preliminary draft. FAO Fish. Techn. Pap., 1969, No. 93, 197 pp., ill. (1970, 9B15).
35. Glover, R.S., Synoptic oceanography. The work of the Edinburgh Oceanographic Laboratory. Underwater Sci. and Technol. J., **2**, No. 1, pp. 34-40, 1970 (1970, 9B48).
36. Grasshoff K., Zur Chemie des Roter Meers und des Interen Golfs von Aden nach Beobachtungen von F.S. *Meteor* während der Indischen Ozean Expedition, 1964-1965. Meteor Forschungsergeben. 1969, **A**, No. 6, 76 pp., ill. (1970, 5B145).
37. Houot, B., Le Bathyscaphe *Archimède* et la recherche oceanographique. Rev. corps santé armées terre, mer, air, **10**, No. 3, pp. 297-304, 1969 (1970, 3B18).
38. Howmiller, R., A horizontal water sampler for investigation of stratified waters. Limnol. and Oceanogr., 14, No. 2, pp. 291-292, 1964, (1970, 1B21).
39. Honot, C.V.G., Le Bathyscaphe *Archimède.* Ann. Inst. océanogr. **46**, No. 1, pp. 5-8, 1968 (1970, 2B16).
40. Hynitzsch, G., Messfahrt in Kuroshio. Seewart, **30**, No. 5, 184-194, 1969 (1970, 9B162).
41. Heller, D., A complete habitat under the sea. Seaports and Shipp. World, 1969, June, pp. 26-27 (1970, 4B10).
42. "Hydroptol"—Elliott hydrop. Marine Res., London, s. a., pp. 22-24.
43. __________. Hudson-70—Canada's first global oceanographic venture. Sea Harvest and Ocean Sci., 1969, June, pp. 60-63 (1970, 2B6).
44. Johnson, F.J., An oceanographic telemetry system. Collect. and Process. Field Data, New York - London - Sydney, 1967, Discuss., pp. 243-248 (1970, 3B26).
45. Jong, E. de., In 600 m Tiefe mit dem Golfstrom. Antrieb. und Steuerung des Tauchbootes durch schwenkbare Motoren. VDI-Nachr. 23, No. 34, p. 9, 1969 (1970, 3B21).
46. Ishino, M., K. Nasu, Y. Morita, and M. Hamada, Oceanographic conditions in the West Pacific Southern Ocean in summer of 1964-1965. J. Tokyo Univ. Fish., **9**, No. 1, pp. 115-208, 1968 (1970, 5B139).
47. __________. *In situ* current speed, current direction, and temperature recording system. Model 501. Underwater Sci. and Technol. J., **1**, No. 2, pp. 97-100, 1969 (1970, 4B53).

48. ________. Die *Komet* arbeitet mit neuen Vermessungsverfahren. Schiffahrt int., 2, No. 9, pp. 4-6, 1969 (1970, 5B18).
49. Laffont, M., and H. G. Delauze, The *Janus* experiment - working at 500 feet. Ocean Ind., 4, No. 2, pp. 52-57, 1969 (1969, 12B13).
50. Lewis, G.L.R., Time for a fresh look at underwater habitat design. Hydrospace, 2, No. 3, pp. 28-32, 1969, (1970, 5B29).
51. Lepetit, J.P., and Mlle. M. Hariguillie, Dépouillement automatique des relevés de fonds sur modèle réduit hydraulique. Bull. Dir. études et rech., A, No. 3, pp. 133-148, 1968 (1970, 1B30).
52. Lee, A.J., *Norwestland* surveys: physical oceanography. Spec. Publ. Internat. Commiss. Northwest Atlant. Fish., 1969, No. 7, Part 1, pp. 31-54, (1970, 5B118).
53. Harms, H., Neuartiges Schiff zur Erforschung des Meeresbodens. Schiff und Hafen, **21**, No. 7, pp. 599-600, 1969 (1970, 2B15).
54. Masuzawa, J., Second Cruise for CSK *Ryofu maru*, January to March 1968. Oceanogr. Mag., **20**, No. 2, pp. 173-185, 1968 (1970, 4B181).
55. ________. Oceanographical observations in the Indian Ocean on the 1964 H.M.A.S. *Diamantina* Cruise DM 4/64. Oceanogr. Cruise Rept., 1969, No. 38, 65 pp., ill. (1970, 8B182).
55a. ________. Oceanographical observations in the Indian Ocean on the 1964 H.M.A.S. *Diamantina* Cruise DM 1/64. *Ibid.*, 1969, No. 3, 74 pp., ill. (1970, 9B161).
55b. Oceanographical observations in the Indian Ocean on the 1965 H.M.A.S. *Diamantina* Cruise DM 2/65. *Ibid.*, 1969, No. 49, 57 pp., ill. (1970, 8B183).
55c. ________. Oceanographical observations in the Indian Ocean on the 1966 H.M.A.S. *Diamantina* Cruise DM 2/66. *Ibid.*, 1969, No. 54, 64 pp., ill. (1970, 8B184).
56. Marcus, S.O., Jr., Evaluation of Navy automatic buoy ocean weather reports. Mariners Weather Log, **13**, No. 4, 152-1955, 1969 (1970, 4B52).
57. Majewski A., Miedzynarodowy Rok Bałtycki, 1969-1970. Gaz. obserw. PIHM, **22**, No. 6, pp. 3-5, 1969 (1970, 2B114).
58. Mason, C.S., and R.L. Gibbert, Some developments in undersea technology at the Bedford Institute. Canad. Aeronaut. and Space J., **15**, No. 3, pp. 84-85, 1969 (1969, 12B24).
59. ________. Integrated Global Ocean Station System (IGOSS). Chronique UGGI, No. 76, 149, 1969 (1969, 12B21).

60. Millard, A.R., Precision measurement of ocean temperatures. Marine Res., London, s.a., pp. 26-27, 32-35 (1970, 5B40).
61. __________. New weather ships on station. Canad. Shipp. and Marine Eng. News, 40, No. 11, pp. 58-59, 72, 1969 (1970, 5B17).
62. Noble, V.E., R.D. Ketchum, and D.B. Ross, Some aspects of remote sensing as applied to oceanography. Proc. IEEE, 57, No. 4, pp. 594-604, 1969 (1970, 2B26).
63. Ogura, Y., A. Takeda, R. Kimura, K. Taira, and T. Nakai, Survey of sea surface temperature of the Tsushima Warm Current with seaborne and airborne radiation thermometers. GARP—severe rainstorm research project. J. Meteorol. Soc. Japan, 47, No. 4, pp. 310-318, 1969 (1970, 6B66)
64. Pritzlaff, J.A., Design characteristics of the *Deepstar* family of vehicles. Paper. Amer. Soc. Mech. Eng., 1968, N WA/UNT - 3, 7 pp., ill. (1970, 1B11).
65. Pidgeon, V.W., Radar, a research tool for oceanography. Ann. N.Y. Acad. Sci., 163, No. 1, pp. 187-193 (1970, 6B58).
66. Pryor, T.A., Commercial test ranges. Oceanol. Int., 4, No. 5, pp. 24-27, 1969 (1970, 7B25).
67. Quick, S.L., An unmanned towed instrument platform for ocean bottom research. Marine Sci. Instrum., New York, 4, pp. 52-69, 1968 (1970, 6B28).
68. Ruff, R.E., Self-contained oceanographic data acquisition buoy. Marine Sci. Instrum., New York, 4, pp. 485-489 (1970, 6B28).
69-70. Romanovsky, V., and A. Le Grand, Cadre polyvalent et bouteilles de prélevements pour l'étude des phenoménes pres du fond. Trav. Centre rech. et études océanogr., 8, No. 3, pp. 5-7, 1969 (1970, 4B58).
71. Schlieben, E.W., SKAMP—an amazing unmanned sailboat! Ocean Ind., 4, No. 9, pp. 38-43, 1969 (1970, 4B5).
72. __________. Hovercraft survey set for Canadian shelf. Oil and Gas J., 67, No. 20, pp. 162-163, 1969 (1969, 12B9).
73. Stewart, R.W., Getting data from the ocean. Collect. and Process. Field Data. New York-London-Sydney, 1967, pp. 191-203. Discuss., pp. 243-248 (1970, 2B12).
74. Smith, P.F., W.S. Richardson, and D. Coburn, A digital tape recording dropsonde, Marine Sci. Instrum., New York, 4, pp. 578-584, 1968 (1970, 6B26).
75. Schick, G.B., J.D. Isaacs, and M.H. Sessions, Autonomous instruments in oceanographic research. Marine Sci. Instrum., New York, 4, pp. 203-230, (1970, 6B24).
76. Somers, M.L., Oceanographic instrumentation and data handling. Underwater Sci. and Technol. J., 1, pp. 22-23, 1969 (1969, 12B23).

78. Tackabery, R. E., Deep Ocean S/T/D measurements. Oceanol. Int., 4, No. 6, pp. 41-43, 1969 (1970, 5B48).
79. Tomczak, M., Jr., Über interne Trägheitsbewegungen in der westliechen Ostsee. Dtsch. hydrogr. Z., 22, No. 4, pp. 158-162, 1969 (1970, 10B87).
80. __________. Unterwasserlabor *Helgoland*. Hansa, 106, No. 15, pp. 1317-1318, 1969 (1970, 2B18).
81. Young, A.W., R.W. Buddemeier, and A.W. Fairhall, A new 60-liter water sampler built from a beer keg. Limnol. and Oceanogr., 14, No. 4, pp. 634-637, 1969 (1970, 6B44).
82. Walden, R.G., A review of oceanographic and meteorological buoy capabilities and effectiveness. Marine Sci. Instrum., New York, 4, pp. 99-105, 1968 (1970, 6B16).
83. Watson, A.G., A spar-buoy for the North Atlantic. Underwater Sci. and Technol. J., 2, No. 1, pp. 23-27, 1970 (1970, 9B21).
84. Warsh, K.L., M. Carstang, and P.L. Crose, A sea-air interaction deep-ocean buoy. J. Mar. Res., 28, No. 1, pp. 99-112, 1970 (1970, 9B24).
85. Wesler, J.E., Buoys as platforms for environmental measurements. J. Hydronaut., 3, No. 3, pp. 132-138, 1969 (1970, 3B23).
86. Wesler, J.E., To instrument the oceans: National Data Buoy Systems. First Annual Offshore Technol. Conf., Houston, Tex., 1969. Preprints. 2, pp. 337-348, Dallas, Tex., 1969, (1970, 11B12).
87. Winn, A.L., D.C. Webb, and K. Burt, Autoprobe, a platform for midwater observation. Marine Sci. Instrum., New York, **4**, pp. 347-353, 1968 (1970, 6B17).